The series publishes short books on topics pertaining to all aspects of the theory and practice of information security, privacy, and trust. In addition to the research topics, the series also solicits lectures on legal, policy, social, business, and economic issues addressed to a technical audience of scientists and engineers. Lectures on significant industry developments by leading practitioners are also solicited.

Synthesis Lectures on Information Security, Privacy, and Trust

Series Editors

Elisa Bertino ⓘ, Purdue University, West Lafayette, IN, USA

Elena Ferrari, University of Insubria, Como, Italy

Elisa Bertino · Sonam Bhardwaj ·
Fabrizio Cicala · Sishuai Gong · Imtiaz Karim ·
Charalampos Katsis · Hyunwoo Lee ·
Adrian Shuai Li · Ashraf Y. Mahgoub

Machine Learning Techniques for Cybersecurity

Elisa Bertino ⓘ
Department of Computer Science
Purdue University
West Lafayette, IN, USA

Sonam Bhardwaj
Department of Computer Science
Purdue University
West Lafayette, IN, USA

Fabrizio Cicala
Department of Computer Science
Purdue University
West Lafayette, IN, USA

Sishuai Gong
Department of Computer Science
Purdue University
West Lafayette, IN, USA

Imtiaz Karim
Department of Computer Science
Purdue University
West Lafayette, IN, USA

Charalampos Katsis
Department of Computer Science
Purdue University
West Lafayette, IN, USA

Hyunwoo Lee
Department of Energy Engineering
(Energy AI Track)
Korea Institute of Energy Technology
(KENTECH)
Naju, Korea (Republic of)

Adrian Shuai Li
Department of Computer Science
Purdue University
West Lafayette, IN, USA

Ashraf Y. Mahgoub
Department of Computer Science
Purdue University
West Lafayette, IN, USA

ISSN 1945-9742   ISSN 1945-9750 (electronic)
Synthesis Lectures on Information Security, Privacy, and Trust
ISBN 978-3-031-28258-4   ISBN 978-3-031-28259-1 (eBook)
https://doi.org/10.1007/978-3-031-28259-1

This Springer imprint is published by the registered company Springer Nature Switzerland AG
The registered company address is: Gewerbestrasse 11, 6330 Cham, Switzerland

Preface

The protection of information and information infrastructures from unauthorized access, use, disclosure, disruption, modification, or destruction is today more critical than ever as they represent attractive targets for a diversity of malicious actors. Those actors have different motivations, such as financial gains, sabotage, espionage, intellectual property theft, and data theft. They may be supported by enemy nations and can leverage an arsenal of attack toolkits, zero-day vulnerabilities, and compromised passwords available on the dark web.

Research and design of defense techniques have, however, greatly progressed over the past 30 years, and there is an increased general awareness of threats and attacks in cyberspace and the need for better defenses by public and private organizations and governmental agencies as well as by the general public.

In order to devise more effective defenses, recent security solutions leverage machine learning techniques, which are today quite effective because of the huge and diversified technical advances in the area of machine learning combined with big data collection and analysis capabilities. We observed that in the last 10 years, the use of machine learning techniques for security tasks has been steadily increasing in research and also in practice. Many recent papers have proposed approaches for specific tasks, such as software security analysis and anomaly detection. However, these approaches differ in many aspects, for example, with respect to the types of features used in machine learning models and the datasets used for training the models. Also, the use of machine learning for security tasks is not trivial. For example, suppose one would like to use machine learning techniques for network intrusion detection. In that case, one has to understand the features to extract from network flows for proper training and use machine learning models able to classify the flows as benign or malicious. To date, however, there is no book or survey article that systematically covers the entire area of machine learning techniques for cybersecurity. This monograph aims to address such a gap.

A comprehensive discussion and analysis of the various machine learning techniques require, however, a proper taxonomy of these techniques. We decided to organize the discussion around the following main cybersecurity functions: security policy learning, software security analysis, hardware security analysis, detection, and attack management.

For some of those functions, many approaches have been proposed, such as the ones for intrusion detection. For others, approaches are still very limited—for example, for attack management. For topics on which many approaches have been proposed, we selected the approaches that we considered most interesting for the discussion. The monograph also covers challenges in using machine learning for cybersecurity—many of which are common to other domains; however, we try, whenever possible, to discuss these challenges from a cybersecurity perspective. Throughout the discussion, we also point out research directions based on our analysis of existing approaches, techniques, and tools. The book also includes a chapter that can be interesting from an educational point of view. This chapter covers three case studies—each related to a well-known cyber attack; for each attack, we discuss which machine learning technique(s) (if any) would have prevented/mitigated which steps of the attack. The case studies are interesting as they show that attacks are typically multi-steps, so one must deploy many different defense techniques to enhance security.

Writing this monograph has been an exciting journey for us as we had several interesting discussions and also identified new ideas for future research. We hope you will enjoy learning about machine learning for cybersecurity as much as we have!!

West Lafayette, IN, USA   Elisa Bertino
West Lafayette, IN, USA   Sonam Bhardwaj
West Lafayette, IN, USA   Fabrizio Cicala
West Lafayette, IN, USA   Sishuai Gong
West Lafayette, IN, USA   Imtiaz Karim
West Lafayette, IN, USA   Charalampos Katsis
Naju, Korea (Republic of)   Hyunwoo Lee
West Lafayette, IN, USA   Adrian Shuai Li
West Lafayette, IN, USA   Ashraf Y. Mahgoub
December 2022

Acknowledgements This work has been partially funded by NSF under Grants DGE-2114680 and CNS-2112471.

Contents

Acronyms

ABAC	Attribute-based Access Control
AI	Artificial Intelligence
ANN	Artificial Neural Network
ARM	Advanced RISC Machine
APT	Advanced Persistent Threats
ARuM	Association Rule Mining
AST	Abstract Syntax Tree
BO	Bayesian Optimization
CVE	Common Vulnerabilities and Exposures
CNN	Convolutional Neural Network
CFG	Control Flow Graph
DA	Domain Adaptation
DDG	Data Dependency Graphs
DGCNN	Deep Graph Convolutional Neural Network
DL	Deep Learning
DM	Data Mining
DNN	Deep Neural Network
DoS	Denial of Service
eBPF	Extended Berkeley Packet Filter
ELF	Executable and Linkable File
FSM	Finite State Machine
GAN	Generative Adversarial Networks
GNN	Graph Neural Network
HI	Hazard Indicator
IoT	Internet of Things
IDS	Intrusion Detection System
IVFG	Inter-procedural Value Flow Graph
LSTM	Long Short-Term Memory
LTL	Linear Temporal Logic
ML	Machine Learning

MM   Memory Management
MOS   Memory Operation Synopsis
MAB   Multi-Armed Bandit
MLP   Multi-Layer Perceptron
NLP   Natural Language Processing
NN   Neural Networks
NSF   Network Security Function
PCA   Principle Component Analysis
PE   Portable Executable
RBAC   Role-based Access Control
RF   Random Forest
RL   Reinforcement Learning
RNN   Recurrent Neural Network
SDN   Software-Defined Networks
SFS   Sequential Forward Search
SGD   Stochastic Gradient Descent
SR   Security Requirement
SR-CR   Security Related Change Request
SVM   Support Vector Machine
SVR   Support Vector Regression
TF-IDF   Term Frequency-Inverse Document Frequency
TL   Transfer Learning
t-SNE   t-Distributed Stochastic Neighbor Embedding
VAE   Variational Auto Encoder
VGG   Visual Geometric Group

Introduction

Attacks to computer, information, and communication systems, collectively referred to as *cyberspace*, are on a dramatic increase. Attacks have a variety of goals, such as data ransoms, denial of service, critical infrastructure sabotage, data theft, and information tampering, and are carried out by many different actors with motivations that include financial gains, cyberwar, misinformation, and disinformation. Security of the cyberspace, referred to as *cybersecurity* (security, for short), is more critical than ever for our society that increasingly relies on cyberspace for all services, functions, and processes we may think of.

It however is well known that there is no system that can be 100% secure from all adversaries. Critical systems, protocols, and software considered secure are constantly analyzed by intelligent adversaries with sufficient resources, leading to the identification of vulnerabilities allowing these adversaries to craft exploits for breaking into computer and network systems. Vulnerabilities, unknown to the creators or users of a system, are called zero-day vulnerabilities, and the exploits that take advantage of them are called zero-day exploits [213]. Recent attacks are increasingly more sophisticated in the vulnerabilities they exploit, and are supported by the availability on the dark web of attack toolkits, detailed zero-day vulnerability information, and compromised security credentials.

The huge expansion of cyberspace due to Internet of Things (IoT) devices and systems, robots, autonomous vehicles, and new wireless and cellular technologies, each with different security postures, has also substantially increased the attack surface. Consequently, the number of adversaries attempting to find new ways of breaking into these systems has skyrocketed. It is clear that protecting the cyberspace requires an array of advanced technical defenses as well as their systematic deployment based on a security life cycle.

In order to devise more effective defenses, recent security solutions leverage machine learning (ML) techniques, which are today widely applied because of their technical significant advances combined with big data collection and analysis capabilities. However, a major problem is that the application of ML techniques to cybersecurity is not trivial. For

© The Author(s), under exclusive license to Springer Nature Switzerland AG 2023
E. Bertino et al., *Machine Learning Techniques for Cybersecurity*, Synthesis Lectures on Information Security, Privacy, and Trust,
https://doi.org/10.1007/978-3-031-28259-1_1

example, if one would like to use ML techniques to classify malware, one has to under-stand the features to extract from malware for properly training and using ML classification models.

1.1   Artificial Intelligence, Machine Learning, and Deep Learning

Today, terms such as artificial intelligence (AI), machine learning (ML), and deep learning (DL) are widely used not only in the technical literature but also in the media, popular culture, advertising, and more [41]. These terms are often used interchangeably. There are however differences among the respective areas that we outline in what follows.

AI is a field that started in 1956. The goal then, as now, was to use computer systems to perform tasks requiring human intelligence [124]. The initial focus was on tasks like playing checkers and solving logic problems. AI then specialized based on specific application areas, such as robotics, natural language processing, and computer vision [13]. Early AI approaches were mainly based on declarative knowledge provided by humans, for example, in terms of logical rules and ontologies. Such knowledge would then be used as input by inference mechanisms, often based on some formal logic. Today, AI encompasses a broad set of technology solutions that can learn on their own.

A major problem of early AI approaches was the lack of scalability because of their reliance on human inputs. ML techniques, which started to be widely used in the1980s, address this problem by relying on data, instead of explicit human input. They apply statistical methodologies to identify patterns occurring in data. They improve their prediction tasks every time they acquire new data. A special category of ML techniques is represented by data mining (DM), which basically addressed the problem of identifying patterns on very large datasets. Most research in DM was initiated by the database community, which for example introduced the pioneering concept of association rule mining [10] and designed efficient algorithms to minimize scans on data stored in secondary storage. However, even though ML techniques can improve their prediction accuracy, "they only explore data based on programmed data feature extraction; that is, they only look at data in the way we program them to do so. They do not adapt on their own to look at data in a different way" [41].

DL techniques represent an important category of ML techniques that address the short-coming of early ML techniques. DL essentially refers to algorithms that adapt, when exposed to different situations or data patterns. Vaguely inspired by biological neural networks, DL algorithms try to learn various characteristics from data and use them for decision-making/prediction on similar unseen data. DL techniques have gained interest because of the increased amounts of data available and their various algorithmic innovations as well as significant improvements in computing capabilities enabled by GPUs, which have made fast training and deployment of DL models possible [169]. DL has been tremendously successful at tasks such as image classification, object detection, and text and voice recognition.

1.2   Security Functions

A detailed and comprehensive discussion of ML-based techniques for cybersecurity is best based on a taxonomy of ML-based *security functions*, that is, security techniques and processes for which ML approaches have been proposed. The taxonomy we refer to is shown in Fig. 1.1. As we can see from the taxonomy, the top five categories correspond to major ML-based security functions that we briefly discuss in what follows.

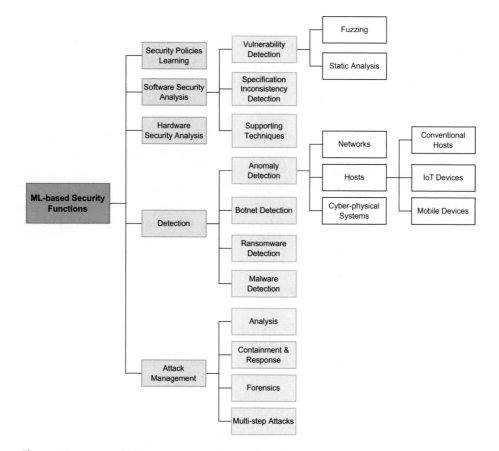

Fig. 1.1 Taxonomy of ML-based security functions

1.2.1   Security Policy Learning

Security policies are critical for configuring security tools and appliances, including access control systems, authentication systems, and network firewalls. As a manual specification of policies is time-consuming and not scalable, it has been one of the first areas to which ML techniques have been applied. The relevance of security policy learning will be increasing given the recent zero-trust architectures [192] and frameworks [133], which will require the specification, deployment, and testing of very large number of policies.

1.2.2   Software Security Analysis

Software systems are key components of all infrastructure and application domains we may think of. However, software systems are still insecure, despite the fact that the "problem of software security" had been known to the industry and research communities for decades [31]. Therefore, it is not surprising that software security analysis has recently become one relevant application area for ML techniques. ML-based approaches range from enhancing fuzzing to ensure better coverage [187] to predicting the effects of different combinations of control parameter values for drones [95] and making static analysis scalable for large code bases [119]. Such initial approaches show that ML techniques can make software security analysis more effective. We can expect that this area will see many novel ML-based approaches to be developed, given the pressing problem of software security.

1.2.3   Hardware Security Analysis

Hardware is commonly assumed to be the root-of-trust for computer systems, in that trust is established by committing functionality to silicon, which represents a stronger security foundation compared to the flexible but more vulnerable software [223]. However, hardware can be attacked, via for example side channels [188], and can even include malicious components (e.g., hardware Trojans). The major use of ML has been for the security evaluation of cipher implementations against side-channel attacks [100] and the construction of attack models against physical unclonable functions [101]. More recent applications of ML include the characterization of faults that can be exploited by attackers [206] and security-aware design flow for chip design [118, 188]. However, ML techniques will undoubtedly enable the design of novel approaches expanding the faults and vulnerabilities that can be detected.

1.2.4   Detection

Detection of security-relevant events, such as intrusions, represents a key security function. Therefore, over the years, many ML techniques have been proposed specifically for assisting intrusion detection systems. For example for misuse-based intrusion detection systems, a ML model can learn from a training dataset containing examples from all of the misuse classes. It can then be used to infer whether a certain event belongs to one of the misuse classes or it is a normal event. In the case of anomaly detection, the model can learn the defined normal behavior and then be used to differentiate between normal and anomalous behavior. Over the years, detection systems have been specialized for different environments (e.g., hosts, networks, cyber-physical systems, and IoT systems) and specific attacks (e.g., ransomware and botnets). As the detection of security-relevant events is critical for comprehensive security, advances in ML techniques will play a critical role in enhancing the accuracy and adaptation capabilities of detection systems.

1.2.5   Attack Management

Properly managing attacks is critical to make sure that the protected system can continue working, perhaps with reduced capabilities, by taking defensive actions for blocking the attack and for recovering. Also, forensics activities are critical for identifying the vulnerabilities exploited by the attack and the attack steps, and for attack attribution. In general, with the exception of forensics analysis, ML techniques have not been used much for attack management. The reason is that the best strategy for attack management is often application-dependent, especially for systems whose continuity, despite being under attack, is critical. However, this is an area where ML techniques can be applied, especially if datasets were available. Also, ML techniques could be critical in the case of multi-step attacks for predicting possible next steps of an attack based on some attack early indicators. Such predictions would be invaluable for containing/stopping the attack.

1.3   Security Life Cycle

The security life cycle is a fundamental notion for systematically organizing security defenses. It typically consists of multiple phases (see Fig. 1.2). As each such phase can be supported by one or more ML-based security functions, in addition to conventional security functions, the security life cycle represents the foundation for a systematic deployment of ML for cybersecurity.

Preparation activities are at the core of security. They include ensuring that messages and data at rest are encrypted, access control systems and firewalls are installed, and applications are free of vulnerabilities. Because of the large number of security tools and techniques

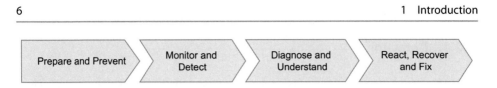

Fig. 1.2 Security life cycle

developed over that security risks be considered when selecting the most effective solutions. In the preparation phase, relevant security functions for which ML techniques are critical include: learning access control policies from data and context; guiding fuzzing techniques to discover application vulnerabilities, ensuring high coverage, and enhancing application testing [187]; identifying where to allocate security resources so to maximize security and minimize its cost as organizations often have limited budgets for security.

However, even the best prepared system can still be breached, and the trade-offs. Therefore, *monitoring activities* are critical for detecting attacks or anomalies that may be indicative of attacks. A real-time anomaly detection system is crucial for enabling quick responses to attacks. Here, ML techniques can be used to build anomaly detection systems, and this is the security function for which several ML techniques have been proposed. At this phase, the dangers of privacy infringements and selective barriers to service also need to be carefully calibrated and explained to the stakeholders. Yet, discrepancies between assumptions and variability in training datasets can create unreliable diagnoses, including false positives and false negatives that can, again, affect stakeholders differently and need to be appropriately communicated. Developing an early and effective ability to think in terms of trade-offs to minimize false signals and maximize accuracy is extremely important.

Once an anomaly or attack has been detected, it is critical to carry out *diagnostic activities* to identify the type of attack, the system components affected by the attack—for example hosts or network links—and depending on the type of attack identifying the affected portions of the system, the entry point of the attack, or the source of attack. Here, approaches based on federated ML techniques and causality reasoning techniques are critical. Yet, discrepancies between assumptions and variability in training datasets can create unreliable diagnoses, including false positives and false negatives that can, again, affect stakeholders differently and need to be appropriately communicated.

Finally, once a diagnosis has been obtained, *response activities* have to be executed. In this phase, ML reinforcement learning techniques are critical to determining the response actions that best contain the attack and minimize the attack damages. As responses entail real-life consequences, from those related to the quality of service and applicability of solutions to differential impact on stakeholders, again, understanding and communicating them is of crucial importance.

These four phases are continuously executed and depending on the situation may even run in parallel. For example, once attacks/anomalies are diagnosed, the prepare phase may be executed again to undertake activities such as patching vulnerabilities—exploited by the attacks—and changing permissions, while at the same time activities are executed to contain the attacks/manage the anomalies.

1.4   Organization of This Monograph

The remainder of this monograph is structured as follows. Chapter 2 provides some background about the main ML techniques used in ML-based security functions and includes pointers to detailed surveys and other resources concerning these techniques. Chapters 3, 4, 5, 6, and 7 cover the ML-based approaches for the main five security functions discussed in Sect. 1.3. In addition to detailed technical discussions, each such chapter includes a discussion on research directions. Chapter 8 presents case studies focusing on well-known multi-step attacks and shows which ML techniques could have mitigated some/all of the steps of these attacks. Chapter 9 discusses a variety of issues in the application of ML techniques to security, including scarcity and quality of training datasets, selection of hyperparameters, models, and configuration, explainability, ethics, and security of models and algorithms. Finally, Chapter 10 outlines some concluding remarks. The monograph also includes an appendix with short descriptions of the publicly available datasets used for experimental evaluations of the ML-based security techniques we have discussed.

Background on Machine Learning Techniques

2

In what follows, we introduce the ML techniques most widely used in security functions. The introduction is meant to provide the high-level concepts of those ML techniques with references to materials providing more in-depth coverage. The chapter also covers transfer learning techniques that are critical to addressing the problem of scarce training data, which is often the case in security. In addition, the chapter covers embedding techniques as they are often used in conjunction with ML.

2.1 Preliminary Notions

An important distinction between the various ML approaches is supervised learning versus unsupervised learning. The main difference is that the former uses labeled data to train the model, while the other does not. Supervised learning is thus a ML approach defined by the use of labeled datasets. These datasets are built to train or "supervise" ML algorithms for accurately executing learning tasks, such as data classification or outcome predictions. They consist of raw data (images, text, videos, etc.) to which one or more meaningful and informative labels are added—typically by humans, to provide semantics and context. For example, labels might indicate whether a photo contains a cat or a vehicle, or which words are present in an audio recording. By contrast, unsupervised learning uses algorithms that discover hidden patterns in data without the need for human intervention, that is, they do not need labeled data.

As the process of labeling datasets can be expensive and time-consuming, a category of approaches, referred to as semi-supervised learning, has been proposed. Such approaches combine a small amount of labeled data with a large amount of unlabeled data. Semi-supervised learning falls between unsupervised learning (with no labeled training data)

E. Bertino et al., *Machine Learning Techniques for Cybersecurity*, Synthesis Lectures on Information Security, Privacy, and Trust,
https://doi.org/10.1007/978-3-031-28259-1_2

and supervised learning (with only labeled training data). Yet another category is represented by the weakly supervised learning; methods in this category learn with coarse-grained labels or inaccurate labels. These methods thus represent another approach for dealing with the problem of obtaining high-quality labeled datasets. More recently, another category of approaches, referred to as self-supervising learning, has been introduced. Such approaches are able to adopt self-defined pseudo-labels as supervision, and use the learned representations for downstream ML tasks. These approaches have been mainly applied to computing vision and natural language processing. We refer the reader to [125, 128] for surveys on self-supervising learning.

2.2   Neural Networks

An artificial neural network (ANN) consists of a pool of simple processing units (neurons) which communicate by sending signals to each other over a large number of weighted connections [137]. The most common form of ANN is the feedforward network, which has a layered structure and where each layer consists of some neuron(s). In a feedforward network, the data always flows from the input layer to the output layer. Each neuron receives inputs from neurons from the previous layer and sends its outputs to neurons in the next layer. There is no data flow within a layer or to the previous layers. In Sect. 2.4, we introduce a different ANN where feedback connections are present. In this section, we focus on feedforward network.

A feedforward network learns a function that maps a set of inputs to a desired set of outputs. The most common learning situation is supervised learning, where the network is trained by providing it with data and matching outputs, that is, a training dataset. During training, the network computes a loss function that measures the distance between the outputs of the network and the desired outputs. The network then changes its internal weights to reduce the distance. As a typical training practice, the network is given many small sets of examples called batches. In each iteration, the network computes the loss for one batch and adjusts the weight accordingly. After training, the network is tested on a new set of data called a test dataset. DL uses multi-layer neural networks to learn representations of data with multiple levels of abstraction [117].

Unsupervised learning is the case where the networks are trained without any labeled training data. These methods discover salient features from the data population. Popular unsupervised learning methods include k-means clustering, agglomerative clustering, and autoencoders (Sect. 2.3).

In a neural network, the weights/parameters are critical features in that they define the input-output function of the network. To understand what those weights are, it is important to have a closer look at neurons again. Each layer of a feedforward network consists of one or more neurons. Figure 2.1 shows a single neuron with two inputs and a single output. Each

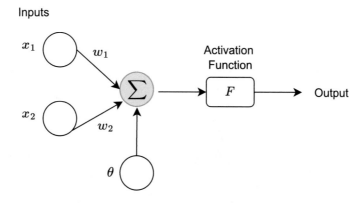

Fig. 2.1 Single neuron with one output and two inputs

connection has a weight w_i where x_i is the input. The final input of the neuron is calculated as the weighted sum of the inputs plus the bias term θ. The output is defined as

$$y = F\left(\sum_i w_i x_i + \theta\right) \tag{2.1}$$

where F is a non-linear activation function. The activation function decides whether a neuron should be activated or not. This means that it will decide whether the neuron's input to the network is important or not in the classification process. The most common non-linear functions include ReLU, Tanh, and Sigmoid. Choosing the most appropriate activation function in the output layer is critical since it defines the type of task the model can perform. We refer the reader to an interesting discussion on how to choose an activation function for neural networks [218].

Today, neural networks often have millions of weights. Every weight can be updated individually to gradually reduce the error over many training iterations. We can think of all these weights as a vector. To adjust such a vector, the learning method computes a gradient vector that contains partial derivatives of the loss function with respect to each weight. The parameter vector is then adjusted in the opposite direction to the gradient vector. This process is repeated in each training step until the loss function stops decreasing. The procedure of repeatedly computing the gradient vector and then updating the parameters is called *gradient descent*.

The critical mathematical process involved in gradient descent is the calculation of the partial derivatives of the loss function with respect to each weight of the network. Backpropagation is an efficient way of computing the gradient vector in a multi-layer neural network via the chain rule of calculus. The backpropagation algorithm works by computing the gradient one layer at a time, iterating backward from the output layer all the way to the input layer. To calculate the gradient at a particular layer, the gradients of all following layers are combined via the chain rule.

2.3   Autoencoders

An autoencoder is a neural network that is trained to reconstruct its input. The network has two components: an encoder e that produces a compressed latent space $h = e(x)$ and a decoder d that produces a reconstruction $\hat{x} = d(h)$. The objective is to minimize the reconstruction error

$$L(x, d(e(x))) = \|x - \hat{x}\|^2 = \|x - d((e(x))\|^2 \tag{2.2}$$

where L is a loss function measuring the distance between x and $d(e(x))$. Autoencoders can be trained with minibatch gradient descent. At each batch, we feed the autoencoder with some data and backpropagate the error through the layers to adjust the weights of the networks.

Autoencoders are used for dimension reduction or feature learning, as they can extract useful information from the data. However, autoencoders can cheat by copying the input to the output without learning useful properties of the data [89]. One way to prevent copying tasks is to have *undercomplete autoencoders*, where the latent space has a smaller dimension than the input. With a reduced dimension, the autoencoders are forced to learn the most important attributes of the data.

While carefully choosing the latent space dimension might avoid that an encoder and a decoder copy the input to the output, a more systematic approach is to enforce the models to have other properties by adding some regularization terms to the loss functions. We now introduce two types of regularized autoencoders: (i) denoising autoencoders, which are robust against corrupted inputs; and (ii) variational autoencoders, which ensure that all the points of the latent space give meaningful content once decoded.

2.3.1   Denoising Autoencoders

Recall that autoencoders minimize the function defined by expression (2.2). A denoising autoencoder instead minimizes the function

$$L(x, d(e(\tilde{x}))) \tag{2.3}$$

where $\tilde{x}$ is x with some added noise. An autoencoder receives corrupted data as input and is trained to reconstruct the original data as its output. The expectation is that the autoencoder learns some useful information about the data in the presence of noise rather than simply copying the inputs.

2.3.2 Variational Autoencoders

Sharp reconstruction of autoencoders is mainly required because of the goal to reconstruct the input perfectly without any constraint on the latent space itself, and the price to pay is that the latent space is likely to have many empty areas. Those empty areas usually give a random/uninterpretable output [191]. Variational Auto Encoders (VAEs) try to solve this problem by regularizing the latent space to be continuous, hence, removing empty areas where they might return meaningless output.

The solution is to output a probability distribution for each latent attribute. For example, if the latent dimension is two, then the encoder will output two distributions for each attribute. The next step is to sample any value from the distribution(s) for the decoder. In VAE, the latent distributions are always Gaussian distributions so that they can be described with the mean and the variance.

This architecture is shown in Fig. 2.2. We denote $N(\mu, \sigma)$ as the latent distribution. The encoder produces a distribution $N(\mu, \sigma)$ which is sampled for a value z. Finally, the sampled point is decoded and the reconstruction error can be computed. The loss function has two parts: a reconstruction term that tends to make the reconstructions as similar to the inputs as possible; and a regularization term that tends to make $N(\mu, \sigma)$ close to a Gaussian distribution. The regularization term uses Kullback-Leibler divergence:

$$L(x, d(e(x))) = \|x - \hat{x}\|^2 + KL[N(\mu, \sigma), N(0, 1)] \tag{2.4}$$
$$= \|x - d((e(z))\|^2 + KL[N(\mu, \sigma), N(0, 1)] \tag{2.5}$$

While VAEs can remove the gaps in the latent space, thus allowing easy random sampling and interpolation, the generated outputs tend to be blurry.

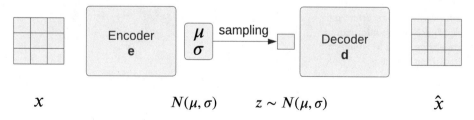

Fig. 2.2 Variational autoencoders

2.4   Recurrent Networks and Long Short-Term Memory

A Recurrent Neural Network (RNN) is a special type of neural network that remembers its previous inputs and outputs. RNNs are very useful in cases where temporal dependencies between inputs are observed, such as handwriting recognition [67], language translation [132], and speech recognition [210]. Due to their ability to *memorize* previous inputs and their corresponding outputs, they are able to process variable-length sequences of inputs.

The main component in a recurrent neuron is the internal memory, see Fig. 2.3), which allows the network to include the *context* when predicting the current output. To show the benefit of the internal memory, consider the task of predicting text sequences. For example, we want to predict the right word to fill the { } in the following sentence: **Mother said *Go play outside*. The boy said *Yes* { }**. Now given the context, the right word can be one of the following: *ma'am*, *mother*, or *mom*. However, if we exclude the context, words such as *sir* or *mister* are also a good fit. In order to assign higher probability scores to the right gender, the network needs to recall the word **Mother** from the previously observed inputs, which is achieved through RNN's internal memory.

In order to leverage the internal memory capability of RNNs, we feed in the inputs serially (e.g., word-by-word) according to their temporal relations. Accordingly, the output of the neural unit for item X_t is included as an additional input when processing item X_{t+1}. However, as we feed the network with new data items, the impact of older decisions on newer items decreases. Accordingly, the network operates as if it has a short-term memory and fails to learn the long-range dependencies across time-steps. This is known as the *vanishing gradient problem* [243]. To address this problem, a few variants of RNN have been introduced. In this section, we describe one popular variant, namely the Long Short-Term Memory (LSTM).

LSTMs solve the vanishing gradient problem by introducing a more complex cell structure, called the memory cell. This key component allows the network to decide what to memorize and what to forget. In other words, instead of trying to memorize the entire history, LSTMs memorize only the relevant items with the highest impact on future items. In

Fig. 2.3 Feedforward neural unit (left) versus recurrent unit (right). RNNs use information from prior inputs to influence the current output

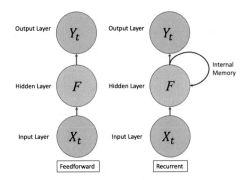

light of our previous example, the word **Mother** is more relevant than words **Go** or **said** in deciding the right gender of the missing word. This information-selective behavior is achieved through the concept of *gates*. A gate is responsible for deciding whether to let information in (remember), or exclude information (forget). This is achieved by multiplying the important indexes in the input vector or matrix by 1, and blocking the non-relevant information by multiplying them by 0. LSTMs use three gates in their structure: (1) Input gate that decides which information to enter in the cell state. (2) Forget gate that tells the cell state which information to keep or forget. (3) Output gate that decides which information to pass to the next hidden state. Note that the difference between cell and hidden state is that the cell state is meant to encode a kind of aggregation of data from all previous time-steps that have been processed, while the hidden state is meant to encode a kind of characterization of the previous time-step's data.

2.5   Attention Mechanism

"Attention" means focusing on or making people notice something. Likewise, the attention mechanism has been introduced in DL to focus on certain factors when processing data. The attention mechanism was initially introduced to address the bottleneck problem in the sequence-to-sequence architecture in neural machine translation. As sentences in a language can be easily modeled as sequences of words, translation from one language to another one can be seen as mapping one sequence to another sequence. Based on this idea, Cho et al. applied deep learning techniques to map two such sequences by using two RNNs [49]. As input sequences can differ in length from output sequences, they introduced a fixed-length data structure referred to as *context vector* in between. Therefore, their approach consists of two steps. In the first step, one RNN compresses a sentence in one language into a context vector. The last hidden state of the RNN is selected as the context vector. Based on the context vector, another RNN generates a translated sentence based on the context vector. The former RNN is an encoder as it is responsible to encode or compress the input sequence into a fixed-length context vector. The latter is a decoder because it outputs the sequence from the context vector. Therefore, we call the translation architecture the *encoder-decoder* approach.

However, Cho et al. [49] pointed out that the encoder-decoder approach has the main drawback due to the context vector. In other words, the encoder compresses all the information of an input sequence into a fixed-length vector. The use of such a vector makes it difficult to handle long sentences, especially for the ones longer than the sentences in the training set. To address such an issue, they designed another neural network that uses all the words in the input sentences instead of only considering the context vector (see Fig. 2.4). Whenever the neural network outputs the next word of the translated sentence, it does not focus on all the words equally. Rather, it tries to identify which parts of the input sequence are relevant to produce the next word accurately. For instance, when a non-English sentence is translated into an English sentence "I did it by myself", it is good to more focus on "I"

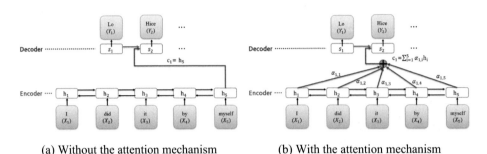

(a) Without the attention mechanism   (b) With the attention mechanism

Fig. 2.4 The architectural difference in the encoder-decoder architecture without and with the attention mechanism in neural machine translation

than other words to generate "myself" (and not "yourself"). As such, an approach pays greater attention to particular input words, the mechanism implementing it is referred to as *attention mechanism*.

The attention mechanism assigns different weights to the input words with regard to the current word in the decoder to generate the most appropriate next word. In detail, given a translated word, the attention mechanism evaluates the *alignment scores* between the translated word and all the hidden states of the input words by using the *alignment score function*. Next, each alignment score is normalized against others, resulting in an *attention weight*. We call a set of all the attention weights with regard to a particular word of the decoder an *attention vector*. Finally, a context vector of the current translated word is calculated as the inner product of the attention vector and the hidden states of the input words. Then, the decoder outputs the next translated words based on the context vector. This process is repeated until the decoder outputs the special word indicating the end of the sentence. By generating the context vector per word (and not one context vector per sentence) explicitly capturing the long-term dependencies, the bottleneck problem is significantly addressed.

Although the attention mechanism was initially introduced in NLP, it can be reformulated into a general form that can be applied to any sequence-to-sequence task where a sequence consists of several *tokens*. To this end, the attention mechanism uses three main components called the *queries Q*, the *keys K*, and the *values V*. A query is a vector attributed to some specific word against the keys with respect to which all the attention scores are calculated. In the above example, the current translated word is a query, and the hidden states of the input words are the keys. The values are a set of tokens scaled according to the attention weights in order to retain focus on the tokens that are relevant to the given query. In the above example, the values are the same as the keys. The attention weights are applied to the hidden states of the input words. However, note that, according to the definition, the values can be different from the keys. Finally, the attention mechanism produces the next token (e.g., the next translated word in the above example).

We refer the reader to surveys [43] for further details on the attention mechanism.

2.6   Reinforcement Learning

RL is a framework by which an agent interacting with a given environment learns an optimal policy, by trial and error, for sequential decision-making problems [145]. By using RL, the agent can acquire near-optimal decision skills, represented by policies, for optimizing user-specified reward functions. The reward function defines *what* the agent should do, and an RL algorithm determines *how* to do [143]. More specifically, a RL framework [220] (see Fig. 2.5) is a probabilistic state transition environment where state transitions are caused by actions executed by an agent and every state-action pair (s, a) is assigned a reward value r given by a reward function $R(s, a)$. The agent traverses the environment according to a policy π that selects an action to execute in a given state for policy parameters θ and receives the rewards accrued over all the state transitions that happened according to θ. In a Q learning framework, the goal is to find the optimal policy which maximizes the cumulative reward through a function $Q(s, a)$ that estimates the expected cumulative reward the agent will get at the end of an episode if the current state is s; the agent executes a and follows learned policy π_θ.

In a deep Q learning system, a deep neural network, referred to as DQN, is used to determine the optimal Q function using a temporal difference equation defined as follows:

$$Q_t(s, a) = Q_{t-1}(s, a) + \alpha[R(s, a) + \gamma Max_{a'}\{Q_t(s', a')\} - Q_{t-1}(s, a)]$$

where Q_t is the current estimation of the Q function, and Q_{t-1} is the previous estimation of Q after $t - 1$ steps of training. The estimated next state and action are denoted by s' and a', respectively. The learning rate (α) determines to what extent newly acquired information overrides old information. The discount factor (γ) determines the importance of future rewards.

One important issue in RL is related to training. There are three main approaches: (i) on-policy, by which the agent interacts with the environment using the latest learned policy, and then uses that experience to improve the policy; (ii) off-policy, by which the agent

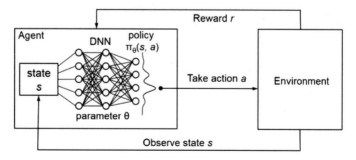

Fig. 2.5 Deep Q learning environment; diagram from [162]

interacts with the environment to collect experience samples but also re-uses samples from older policies, thus enhancing efficiency; and (iii) offline, by which the agent is trained on a static dataset of fixed interactions and thus does not have to interact with the environment. Offline training is used when interactions with the environment are difficult, expensive, or dangerous and has thus generated considerable interest [143]. However, it requires suitable training data. Additional approaches are however also possible by which the agent can be pre-trained in a constrained/simulation environment and then adapted to a real-world environment with a few actual interactions with the environment [144], or an agent trained in one environment can be adapted to a new environment, related to the initial one, by using transfer learning [254].

Another critical issue is related to safety, which refers to complying with safety constraints during the training and/or deployment processes [83]. Approaches proposed for safe RL can be classified into two categories. The first approach is based on modifications to the reward function with the introduction of a safety factor. The second approach is based on the modification of the exploration process with the incorporation of external knowledge. A concrete example of the latter can be found in the context of smart IoT systems [161]. We note that similar approaches may have to be applied when security is a requirement for the application domain of interest, in that one must make sure that decisions taken by an RL agent do not introduce security risks.

We refer the reader to well-known surveys and books [134, 145, 220, 245] for further details on RL techniques.

2.7 Transfer Learning

Transfer learning (TL) refers to the ML problem of using knowledge gained while solving one learning task and applying it to a different but related learning task. In the real world, we observe many TL examples. We may find, for example, that learning audio data from many speakers may help understand the audio data of a specific user. The key observation is that humans can apply knowledge learned from previous tasks to solve new problems faster. In ML, the ability to transfer the knowledge gained by learning one or more source tasks in a domain, referred to as *source domain*, comes into play when dealing with limited training data in another domain, referred to as *target domain*.

2.7.1 Notations and Definitions

We follow the definitions by Pan and Yang [176]. A domain $\mathcal{D}$ consists of a feature space $\mathcal{X}$ and a marginal probability distribution $P(X)$, where $X = \{x_1, \ldots, x_n\} \in \mathcal{X}$. Given a specific domain $\mathcal{D} = \{\mathcal{X}, P(X)\}$, a task $\mathcal{T}$ consists of a label space $\mathcal{Y}$ and an objective predictive function $f(\cdot)$, which can also be viewed as a conditional probability distribution $P(Y|X)$. In general, we can learn $P(Y|X)$ in a supervised manner from the labeled data $\{x_i, y_i\}$, where $x_i \in X$ and $y_i \in Y$.

Assume that we have two domains: the dataset with sufficient labeled data is the source domain $\mathcal{D}^s = \{X^s, P(X)^s\}$, and the dataset with a small amount of labeled data is the target domain $\mathcal{D}^t = \{X^t, P(X)^t\}$. Each domain has its own task: the source task is $\mathcal{T}^s = \{Y^s, P(Y^s|X^s)\}$, and the target task is $\mathcal{T}^t = \{Y^t, P(Y^t|X^t)\}$. In traditional deep learning, $P(Y^s|X^s)$ can be learned from the source labeled data $\{x_i^s, y_i^s\}$, while $P(Y^t|X^t)$ can be learned from labeled target data $\{x_i^t, y_i^t\}$. Here, we give a formal definition of TL from Pan et al. [176].

Definition 2.1 (*Transfer Learning.*) Given a source domain $\mathcal{D}^s$ and learning task $\mathcal{T}^s$, a target domain $\mathcal{D}^t$ and learning task $\mathcal{T}^t$, transfer learning aims to help improve the learning of the target predictive function $f_t(\cdot)$ in $\mathcal{D}^t$ using the knowledge in $\mathcal{D}^s$ and $\mathcal{T}^s$, where $\mathcal{D}^s \neq \mathcal{D}^t$ or $\mathcal{T}^s \neq \mathcal{T}^t$.

2.7.2 Fine-Tuning

While traditional ML techniques learn a model from scratch, the fine-tuning approach makes use of a pre-trained model that is already trained by another dataset. The model uses labeled target data as a guide for transferring knowledge between different domains. The size of the output layer is usually modified based on the label size of the target task. Several layers preceding the output layer are set to be trainable while the earlier n layers are frozen. The weight of the trainable layers is updated using target data to minimize errors between predicted labels and true labels. The first n layers can also be fine-tuned depending on the size of the target dataset and its similarity to the source dataset [239]. The fine-tuning approach has been applied to different tasks, including machine fault diagnosis [208] and network intrusion detection [214].

2.7.3 Domain Adaptation

Based on the definition of TL, the domain shift can be caused by domain divergence $\mathcal{D}^s \neq \mathcal{D}^t$ or task divergence $\mathcal{T}^s \neq \mathcal{T}^t$. Domain adaptation (DA) is the case where the source task $\mathcal{T}^s$ and the target task $\mathcal{T}^s$ are the same, and the domains are related but different. When the domains are different, then either ❶ the feature spaces are different (heterogeneous DA), i.e., $X^s \neq X^t$, or ❷ the feature spaces are the same for both domains but the probability distributions are different (homogeneous DA), i.e., $P(X)^s \neq P(X)^t$. In the example of host-based malware detection, Case 1 corresponds to when the two datasets represent two hosts with different underlying operating systems. Case 2 corresponds to when the source and target datasets are from the same host, but they focus on different attack families.

2.7.3.1 Adversarial Domain Adaptation

We introduce a DA approach referred to as adversarial DA, which typically leverages the generative adversarial networks (GANs) [88]. The idea is to use a domain discriminator to encourage domain confusion through an adversarial objective.

Historically, the use of GANs focused on generating data from noise. Its main goal is to learn the data distribution and then create adversarial examples that have a similar distribution, to deceive an image classifier. However, the use of GANs extends beyond the recognition domain and GANs have been applied to generate benign data, attack data, or both, in the area of cybersecurity.

A basic GAN consists of a generative model, called generator G, and a discriminative model, called discriminator D. The generator G generates data that are indistinguishable from the training data and the discriminator D distinguishes whether a sample is from the data generated by G or from the training data by predicting a binary label. The training of the GAN is modeled as a minimax game where G and D are trained simultaneously and get better at their respective goals: training G to minimize the loss in Eq. 2.6 while training D to maximize it:

$$\min_{G} \max_{D} V(G, D) = E_x[log D(x)] + E_z[log(1 - D(G(z)))] \tag{2.6}$$

where E_x is the expected value over all real instances, E_z is the expected value over all the generated data instances, $D(x)$ is the probability of D predicting a real instance as real, and $D(G(z))$ is the probability of D predicting a generated instance as real.

In adversarial DA, this principle has been employed to ensure that the network cannot distinguish between the source and target domains. The key to adversarial DA is learning a domain-invariant representation from source and target datasets. A good domain invariant representation can be directly used to train a classifier that performs well on both domains.

In the approach by Singla et al. [215], the DA GAN architecture integrates a classifier to ensure that the representation is domain-invariant and meaningful in terms of classification. The discriminator is trained to minimize the loss in Eq. 2.7, where E_{x_s} and E_{x_t} are expected values of the source and target samples, $D(G(x_s))$ is the probability of predicting a source domain sample as belonging to source, and $D(G(x_t))$ is the probability of predicting a target domain as belonging to the source. The classifier is trained to minimize a binary cross-entropy loss—Eq. 2.8. The approach by Singla et al. [215] focuses on the intrusion detection problem where the data is either labeled as attack or benign. $E_{x_{attack}}$ and $E_{x_{benign}}$ are the expected values of attack and benign samples, $C(G(x))$ is the probability of the classifier predicting a sample as attack and $1 - C(G(x))$ is the probability of classifying a sample as benign. The generator has two objectives, maximize the domain classifying loss and minimize the classification loss. Hence, it is trained using both Eqs. 2.9 and 2.8:

$$L_d = -E_{x_s}[log\, D(G(x_s))] - E_{x_t}[log(1 - D(G(x_t)))] \tag{2.7}$$

$$L_c = -E_{x_{attack}}[log\, C(G(x))] - E_{x_{benign}}[log(1 - C(G(x)))] \tag{2.8}$$

$$L_g = -E_{x_t}[log\, D(G(x_t))] \tag{2.9}$$

To apply this approach in a heterogeneous DA setting where the source and target datasets can have different dimensions, DA GAN transforms both datasets to have the same dimension using the Principal Component Analysis (PCA).

2.8 Embedding Techniques

The idea of representing classifiable items as vectors (a.k.a. embeddings) has been investigated in many domains. Examples of these items are words, sentences, and paragraphs in the NLP domain, or malicious and benign applications in cybersecurity. By representing the items as vectors (i.e., transforming them into vector space), multiple operations can be applied such as computing Euclidean or cosine distance, or more complex operations such as ML classification or clustering. Accordingly, the vector representation of an item should preserve its semantics, allowing for similarity and composition operations to be performed accurately.

Different techniques for vector space representation have been proposed. The simplest technique is to manually identify the semantic features of an item and assemble them into a vector representation. This technique is very similar to manual feature engineering, and hence suffers from the same challenges such as problem instance dependency or handling large datasets. Furthermore, it is very hard to apply in domains where features are not easily extracted based on domain knowledge such as image classification.

Due to the shortcomings of manual embedding techniques, automated techniques that are easier, more generic, and less prone to errors have been proposed. The most common automated embedding technique is the autoencoder. An autoencoder (see also Sect. 2.3) is a type of a neural network in which the input and the output are the same, with a hidden layer of reduced dimensionality. Mapping the input to the hidden layer serves as an encoding step, whereas mapping the hidden layer to the output (same as input) serves as a *decoding* step. If the output can be accurately reconstructed from the hidden layer's reduced dimensionality, then the hidden layer representation can be used as a semantic-preserving embedding for this input. We refer the reader to surveys by Goyal and Ferrara [91] on graph embedding and by Almeida and Xexeo [6] on word embedding, respectively.

Security Policy Learning

Security policies define a set of rules and procedures that must be followed in order to protect assets and users from insider or outsider threats. An important category of policies is represented by access control policies, by which policy developers and security officers specify how users may interact with resources. This implies defining varying groups of users with different sets of permissions on resources.

Access control policies are expressed based on access control models such as Attribute-based (ABAC) [102] and Role-based access control (RBAC) [199]. The policy is defined in terms of user characteristics (e.g., attributes or roles), resources, and operations. For example, all users with the *Engineer* role are permitted "read" and "write" operations on the *Customers* database. On the other hand, the policy can be explicitly expressed based on low-level information, such as login identifiers and IP addresses. For example, Alice's laptop with IP address 192.168.5.2 is permitted to send HTTP traffic to the web server with IP address 172.50.2.58, or all traffic originating from VLAN 15 can be routed to the Internet. In both those examples, the policy specification is time-consuming and error-prone [122, 133].

On the other hand, when using a high-level policy model, one must have detailed semantic knowledge, such as roles and attributes. At the network level, network administrators must have a thorough knowledge of the network entities, such as IP/MAC addresses, protocols used, involved ports, and allowed or denied communications. Even the slightest error in the specification could result in unintended accesses or legitimate communication failures.

Another important category of security policies is privacy policies. Privacy policies are the basis with respect to which (i) organizations deploy security controls and privacy mechanisms; and (ii) inform end-users about privacy choices the users have when interacting with the organizations and the use of their data by the organization. Those policies are typically specified as legal documents in natural language. One major issue with these policies is that they can contain contradictions [16], potentially leading to various problems. First, the conditions and actions of the policy become unclear due to such contradictions. For instance,

E. Bertino et al., *Machine Learning Techniques for Cybersecurity*, Synthesis Lectures on Information Security, Privacy, and Trust, https://doi.org/10.1007/978-3-031-28259-1_3

a policy part may allow data, which is privacy-sensitive, to be transmitted via untrusted channels, whereas another part of the policy may prohibit it. Second, if the privacy policy is erroneous, it is extremely hard to investigate whether deployed software systems comply with the privacy policy.

Therefore, approaches have been proposed for automatically learning security policies using ML techniques. Such approaches aim to reduce the number of errors due to the manual specification of policies and create more intelligent access control systems that can adapt to policy changes. However, the design of such ML-based approaches must address specific requirements [28] to ensure that all legitimate accesses and actions are granted, and all illegitimate accesses and actions are denied. We define those requirements as follows:

1. *Policy completeness*: It means that the ML-based learning system should be able to make a decision on any received access request.
2. *Policy correctness*: It means that the ML-based learning system should be able to make the correct decision for any received access request.

In what follows, we first discuss learning the access control policies from data while also discussing approaches to transfer and adapt the learned policies to other domains. Then we discuss DL approaches to policy learning and conclude the section with model-independent policy learning. Then we focus on network security policies as these policies are very critical today. First, we discuss rule mining techniques for firewalls and present approaches for training ML models for firewall systems. Then we cover ML techniques for learning the security policy for traditional and IoT networks. Then we discuss contradiction identification in privacy policies. Finally, we discuss adaptive policy learning systems and conclude the chapter with research directions.

3.1   Access Control Policies

Access control policies are typically composed of conjunctions of predicates depending on the security model. ABAC and RBAC are two widely adopted and well-studied security models. The security model formalizes the specification of security policies and allows the proof of security properties [198].

Example

The following is an example of an ABAC rule expression.

```
<Rule10: user (ID: pr-1, business_unit: marketing),
    resource (ID: strategic_planning_document,
        business_unit: marketing),
    edit, permit>
```

"Rule10" is a unique identifier for the rule. "user (ID: pr-1, business_unit: marketing)" refers to some user with identifier "pr-1" appointed to the "marketing" business unit. The resource in this example is a document called "strategic_planning_document," and the aforementioned user is permitted edit access.

3.1.1 Learning Access Control Policies

Several approaches have been proposed that use ML techniques to learn (or mine) access control policies from data, such as logs of access requests along with the decisions (e.g., allow or deny). A recent approach is *Polisma* [122], a framework that uses data mining, statistical, and ML techniques to learn ABAC policies using access control logs and contextual information. Such contextual information can be obtained from multiple data sources, for instance, LDAP directories, SIEM systems, and organizational charts. The framework comprises the set of users and their attributes, resources and their attributes, operations, and attribute relations. Polisma organizes the ABAC policy learning process into four steps (see Fig. 3.1).

(1) Rule mining: Polisma uses Association Rule Mining (ARuM) [9] to analyze the association between users and resources and extract the rules from the examples provided. In particular, Polisma uses the Apriori ARuM [10] algorithm to generate the rules. However, the rules resulting from rule mining are potentially overfitted or unsafely generalized. In the former case, the rules have attributes that are not useful for making decisions. Thus, overfitting impacts the system's completeness. In the latter case, the rules are too generic and thus could give access to a set of users larger than the intended one. Therefore, unsafe generalization impacts the correctness of the system.

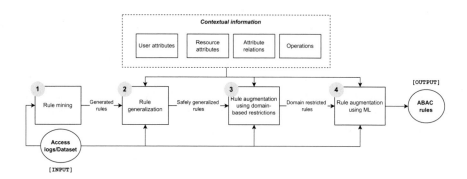

Fig. 3.1 Polisma architecture [122]

Example

The following is an example of two rules generated using ARuM. The first is an overfitted rule, while the second is an unsafely generalized rule.

```
<Rule1: user (ID: developer-1), resource (ID: git-1), read,
   permit>
<Rule2: user (group: engineers), resource (ID: git-1), write,
   permit>
```

(2) Rule generalization: Polisma provides two strategies to address the unsafe generalization and overfitting issues. The first is the brute force strategy, where each generated rule is post-processed by analyzing the user and resource attributes and statistically choosing the ones that grant permission to specific subsets of users on the resource. The second strategy assumes that attribute-relationship metadata (i.e., attribute relations in Fig. 3.1) is given to the system. Thus, Polisma constructs a graph out of those relations. Then, it traverses the graph and selects the minimum set of attributes of users and resources that grant permission to access the resource.

(3) Rule augmentation using domain-based restrictions: The previous steps focus on generating rules that grant permission for accessing resources (i.e., positive/allow rules). However, to address the completeness requirement, the system applies this step to generate negative rules (i.e., deny rules). The system identifies the attributes in the dataset or the contextual information to evenly partition the users and resource sets. Then, it generates negative rules for a user group to perform a specific operation on a resource or a group of resources.

(4) Rule augmentation using ML: While the previous steps generate rules based on the knowledge extracted from the training dataset and contextual information (if available), these rules cannot handle new requests similar to old ones. Specifically, the new request's access patterns are similar to the old ones, but the new request's attributes do not match the old ones. Suppose that software engineer A from department X has access to the git server in the same department. Now imagine a software engineer B from department Y requesting access to the git server in department Y. Clearly, the access pattern is very similar. Polisma trains an ML classifier (e.g., Random Forest and kNN) on the attributes provided in the training dataset and context information. Then the classifier predicts the decision on a new request based on the similarity to the rules generated in the previous steps.

The performance of Polisma has been evaluated using two metrics: (i) the F1 metric; and (ii) the percentage of covered requests (PCR), which assess the policy completeness. The evaluation has been carried out on the Amazon Access Samples Dataset and the Project Management Dataset (see Appendix). The results show that Polisma achieves an F1 score of

0.98 and a PCR of 1.00 when using the Amazon Access Samples Dataset, and an F1 score of 0.86 and a PCR of 0.95 when using the Project Management Dataset.

Das et al. [62] proposed an algorithm to mine minimal sets of ABAC rules that consider environmental variables. Environmental variables typically complement the policies by adding environmental conditions that must be met to approve an access request. For instance, a software engineer may access a production server only if they are located within the company's physical premises.

The proposed algorithm takes as input the sets of subjects, objects, and environmental variables. It also expects a three-dimensional matrix, where the first dimension represents the subjects (s), the second the objects (o), and the third the environmental variables (e). If the value at the point (s, o, e) is 1, the access should be granted; otherwise, it should be denied. The algorithm's output has three objectives: (1) Each entry in the matrix should be covered by at least one tuple in the mined policy, (2) no permissive rules should exist, and (3) the number of rules in the mined policy should be minimum.

The algorithm first computes the *gini impurity*[1] [36] for each attribute data and selects the attributes that best divide the accesses into a permit or deny decisions. The less the gini impurity, the better the access split. Then it recursively repeats the process following the different gini impurities until the algorithm obtains groups of accesses leading to a single decision. Refer to Sect. 3.2 of the paper for more details on the formula used to compute the gini impurity. The authors demonstrate that this approach generates as many or fewer rules as other proposed methods.

Molloy et al. [159] have proposed *generative models* for mining usable and accurate RBAC policies. While accuracy deals with the precision of the generated RBAC policy (i.e., whether roles and permissions have been correctly assigned to the users), usability focuses on mining practically usable models based on the user's usage of the permissions (i.e., entitlements). For example, if two users have the same permissions but different usage of those permissions, they should be assigned to different roles in the generated model. An additional advantage of the generative models is that they are explainable in the sense that there is a reason why a role is assigned to a user. The authors also define a variation of such models called *attributive generative models*. This variation states that roles assigned to the users are causally correlated with the user attributes (e.g., affiliated department), and the assignment of permissions to roles is based on usage patterns.

The author's approach is based on two ML models:

(1) Latent Dirichlet Allocation (LDA): LDA [34] is a probabilistic generative model that works with collections of documents composed of discrete data. Each document is a finite mixture of topics, and each topic represents a probabilistic distribution over words. A document (which is a set of words) could be assigned to multiple topics. In the role mining task, each document represents a user's observed actions, and the extracted topics correspond

---

[1] Gini impurity is a measurement method used to determine how nodes should be split in a decision tree.

to roles. Thus, LDA estimates the roles (topics) assigned to each user's observed actions (document).

(2) Author-Topic Model (ATM): ATM [193, 194] is an extension of LDA modeling reasoning about the documents and the corresponding authors. An author is a multinomial distribution over the topics, and a topic is a probability distribution over the words of the documents, as mentioned above. In the task of role mining, an author models a user with explicit attribution.

In summary, the authors represent the user attributes, the documents represent the user actions and usage, and the topics represent the observed roles. The model learns to extract the roles based on the observed users' actions and automatically assigns users to the appropriate roles. We refer the interested reader to [159] for technical details on the model parameters and algorithms.

Molloy et al. have evaluated their approach to real-life data, such as access logs and entitlement data. The approach exhibits a good performance as it produces a stable decomposition in less than one hour for a dataset of 36M actions performed by 2050 users. Thus, the most significant advantage of the approach is the dramatic performance improvements compared to other techniques. In addition, the approach generates only a tiny amount of over-assignments (i.e., users are assigned to more roles than they should) and shows a low percentage of model uncoverage (i.e., permissions that have not been given to users that thus may not be able to complete all their tasks).

Karimi et al. [131] proposed an approach to detect patterns in access records using unsupervised learning and extract the ABAC policy from those patterns. Each line in the access record represents a tuple composed of the user, resource, session attributes, and access rights. It is important that the authors only consider positive permissions (i.e., allow rules only). The clustering algorithm maps each tuple to a cluster, and each cluster's centroid represents a policy rule. The distribution of the attribute values of each tuple appointed to a cluster and the centroids' attributes are used to extract the policy rules. The steps are as follows.

(1) Data Selection and Preprocessing: In this step, the algorithm parses the data files and collects the various features and their values. In cases of missing values, the algorithm uses the special value "UKN" (standing for "unknown").

(2) Tuning Learning Parameters: In this step, the algorithm discovers (i) which should be the number of clusters (k), (ii) which learning algorithm to use, and (iii) the cluster initialization parameters and local optima. This step uses two methods to select the best value for k. First, it uses the elbow method [92], which is the within-cluster sum of squares (WCSS). In this method, one steadily increases k and observes how the value of WCSS decreases. The point for which the decrease is negligible is a good value for k. In addition to this method, this step computes the mined model's accuracy and selects the k that gives the highest accuracy in a tenfold cross-validation. The accuracy is defined as follows:

$$Accuracy = \frac{TP + TN}{TP + TN + FP + FN} \tag{3.1}$$

where

- TP: The outcome of the access request is allowed both in the original and mined policies.
- TN: The outcome of the access request is denied both in the original and mined policies.
- FP: The outcome of the access request is denied in the original policy, but it's allowed in the mined policy.
- FN: The outcome of the access request is allowed in the original policy, but it's denied in the mined policy.

The authors found that *k-modes* [107], a variation of *k-means* [127], is best suitable for this task due to the use of categorical features. Multiple models are trained with different cluster initialization, and the model with the smallest clustering error is selected.

(3) Rule Extraction: This step first performs clustering with all the available features, such as user and resource attributes. The respective rule is extracted from the attribute values of the clusters' centroid. However, it is possible that not all of those attributes are really effective in the mined policy. To check the effectiveness of an attribute in the extracted rule, the occurrence frequency of the attribute's value is measured in the original dataset and within the cluster. If the difference is higher than a pre-set threshold value, d, the attribute and its value are deemed effective and are part of the extracted rule.

Example

Table 3.1 shows an example of the extracted attributes from a cluster's centroid. The attribute *type* has the discrete values *application, gradebook, roster*, and *transcript* based on some university dataset. Suppose that *type = application* occurs in 60% of the rules in the cluster and in 25% of the rules in the dataset. If $d = 30\%$, then *type = application* is effective and will be part of the rule generated from the cluster.

Table 3.1 An example of a cluster centroid (adapted from [131])

User_department	Object_department	Position	Type	Location
CS	CS	Staff	Application	Campus

(4) Policy Improvement: The step aims to improve the quality of the mined policy using two algorithms: rule pruning and policy refinement. Those algorithms aim to eliminate rules that are subsets of one another and refine the policy to reduce the number of FN and FP.

Karimi et al. evaluated their approach using synthetic datasets with positive permissions. Thus, only examples with allowed permission are given as input to the unsupervised learning model. The evaluation shows that their approach can achieve an accuracy of 99% after policy refinements.

3.1.2 Policy Transfers Across Domains

Jabal et al. also proposed FLAP [123], an extension of Polisma, that focuses on transferring ABAC policies from a source domain to a target domain. Such a transfer is useful when the target domain has limited training data, whereas the source has a sufficiently large training dataset. Each domain has its own datasets, such as logs or existing policies, each perhaps with different characteristics. The main issue in policy transfer is handling conflicts between the policies of the source and target domains. Such conflicts could arise, for example, due to regulatory or policy differences.

The FLAP approach consists of the following steps:

(1) Rule Similarity Analysis: In this step, FLAP evaluates the similarity of each rule from the source domain to the log of the access control decision examples or rules at the target domain. The similarity process checks (a) whether the user in the example satisfies the user attribute expression in the source domain rule; (b) whether the resource in the example satisfies the resource attribute expression in the source domain rule; and (c) whether the operation in the example is included in the source domain rule's operation set. A consistency check is also performed to verify whether a source domain rule is similar to an access control decision example in the target domain. If the source domain rule has the same decision as the example in the log, then they are deemed consistent; otherwise, they are not.

(2) Rules Adaptation: This step applies to inconsistent rules found in the previous step. Given two inconsistent rules, FLAP identifies the mutual and non-mutual attributes in the rule expressions. The proposed algorithm first derives rules based on the mutual attributes while setting the decision to either permit or deny. The access decision is set based on the number of conflicting decision examples. Then it derives new rules based on non-mutual attributes (see Algorithm 1 in [123]).

The FLAP policy transfer process has been evaluated according to two approaches:

(1) Policies Transfer using Local Learning: The rule generated after each Polisma step (see Fig. 3.1) is compared against the source domain rules using the rule similarity analysis. If there is a conflict, the local rules are adapted accordingly using the rule adaptation while conflicting source domain rules are filtered out. The remaining source domain rules after all the steps are transferred to the target domain.

(2) Policies Transfer using Hybrid Learning: In this method, the source domain rules are combined with the intermediate local rules allowing Polisma to leverage both the source and target domains' knowledge for the next step. The intermediate rules are adapted based on the source domain rules, while the conflicting source domain rules are discarded.

The FLAP transfer learning approach was evaluated using the Amazon Access Samples Dataset and the Project Management Dataset (see Appendix). The results showed that the generated rules achieved an F1 score of about 0.80 on the Amazon Access Samples Dataset and an F1 score of about 0.75 on the Project Management Dataset.

3.1.3 DL Models for Access Control Decisions

The previous approaches take input data, such as logs, histories of accesses, and access control lists, and employ ML techniques to generate ABAC rules as output. A different approach is by Nobi et al. [172], which proposes an approach that generates a trained DNN instead of access control rules.

During training, the DNN receives metadata and access requests as input and decides whether access should be permitted or denied. Metadata information could be any relevant information including access logs, employee join date, access times, etc. Access request tuples are in the form $<user, resource, operations>$. The DNN has a classification layer where the number of neurons equals the number of operations. Each neuron outputs the probability of permitting or denying the execution of a related operation to a user. The DNN is mathematically abstracted as $\hat{y} = f(x)$, where $\hat{y}$ is the access prediction probability for executing a certain operation, and x is the user and resource metadata. If the prediction is greater than a threshold (0.5, in the paper), access is permitted; otherwise, it is denied. Since the samples in the training set have categorical features, the feature values are transformed into a two-dimensional array using one-hot encoding.[2] The rows represent training examples, and each column holds the value of each feature (that is, metadata values).

Once the DNN is trained, it is used for the access decision-making process. Specifically, the system receives the tuple $<user, resource, operation>$ and fetches additional metadata information from internal data stores. Then, the DNN decides whether the request should be approved or not. The authors have demonstrated the effectiveness of the proposed approach using three DNN architectures: ResNet [99], DenseNet [106], and Xception [59]. They used two real-world datasets (dataset #1 and #4 in Appendix) and eight synthetic datasets (dataset #3 Appendix). The results show that the F1 score ranges from 0.75 to 0.95 depending on the dataset.

---

[2] Some ML algorithms require categorical features to be converted to a numerical form. If the categories are independent (e.g., "cat" and "dog"), then the one-hot encoding breaks and replaces the feature with one binary feature per category. For example, if a feature can only have two labels, "cat" and "dog", it will be converted to the vector [1,0] if the feature has the label "cat" or [0,1] if it has the label "dog".

3.1.4 Model-Independent Policy Mining

One limitation of the existing policy miners, such as [122, 197, 248], is that their design and implementation are based on specific access control models. Thus, the miners must be updated depending on the access control models they need to support. Cotrini et al. [56] proposed Unicorn, a universal method for access control policy mining. Due to its generic design, Unicorn generates policy miners for diverse policy models, such as ABAC and RBAC with spatial-temporal constraints [4] without assuming any prerequisite in ML techniques and combinatorial algorithms.

In order to apply Unicorn, one needs to perform three steps:

(1) Template: In this step, the user must express the model as a first-order logic formula defined as $\varphi_b(u)$, where u is a user in the organization. The template formula contains Boolean variables whose assignment must correspond to a policy.

Example

Consider the following ABAC rules:

<Alice (position = "Doctor"), resource (ID = "databzase1"), 1>
<Bob (position = "Doctor"), resource (ID = "database1"), 1>
<Nick (position = "Nurse"), resource (ID = "database1"), 0>
<Nick (position = "Secretary"), resource (ID = "database1"), 0>

The convention followed in those rules is <user (attribute = value), resource (attribute = value), 1—Allow/0—Deny>. A potential template formula for those rules is the following:

$$\varphi_b(u) = (attVal(u) = Doctor \wedge b_1) \vee (attVal(u) = Secretary \wedge b_2) \vee (attVal(u) = Nurse \wedge b_3)$$

where b_1, b_2, b_3 are Boolean variables, u is a user in the organization, and $attVal(u)$ is a function that maps users to their attribute values (i.e., the attribute "position" in this example).

For $b = (0, 1, 1) = b_1 = 0, b_2 = 1, b_3 = 1$, we have
$$\varphi_{(0,1,1)}(u) = (attVal(u) = Doctor \wedge 0) \vee (attVal(u) = Secretary \wedge 1) \vee (attVal(u) = Nurse \wedge 1) \Leftrightarrow$$
$$\Leftrightarrow \varphi_{(0,1,1)}(u) = (attVal(u) = Secretary) \vee (attVal(u) = Nurse) =$$
$$= \{Secretary, Nurse\}$$

The above statement defines a policy for secretaries and nurses. For $b = (1, 0, 0)$, we would have $\varphi_{(1,0,0)}(u) = Doctor$, and this is a policy that applies to doctors.

The above example shows that the template formula may express an arbitrary policy independent of the access control model of choice. The example also shows two *interpretations* of the template formula; $b = (0, 1, 1)$ and $b = (1, 0, 0)$. Thus, the policies are nothing more than interpretations of the user-defined template formula. Hence, mining the policy becomes a searching problem in the interpretation space.

(2) Objective Function: An objective function measures how well the generated policy fits the log.

Example

A simple example of an objective function is

$$L(b) = \sum_u |Log(u) - \varphi_b(u)|$$

where $Log(u) : \{1, 0\}$ is a function that returns whether the user receives permission for particular access based on the log, and $\varphi_b(u)$ is the decision result from the template function. $L(b)$ is, in this case, the Hamming distance between $Log(u)$ and $\varphi_b(u)$ given an interpretation b.

Another possible objective function is

$$L(b) = \lambda \sum_u |Log(u) - \varphi_b(u)| + Complexity(\varphi_b)$$

Here, the sum is multiplied by a constant weight λ. *Complexity* is a user-specified function that quantifies the complexity of the template formula (e.g., number of predicates).

(3) Implementation of Unicorn's Pseudocode: In this step, Unicorn attempts to generate a policy that minimizes the objective function using Unicorn's pseudocode. The approach used to mine the policy is *deterministic annealing with mean-field approximation* [56]. Deterministic anneal is a greedy algorithm that tries to minimize $L(b)$. However, for template functions with several Boolean variables, the problem becomes intractable. Hence, approximations are used to find a good solution to the problem.

Unicorn has been evaluated based on three metrics: True Positive rate (TPR), False Positive Rate (FPR), and Complexity. TPR expresses the fraction of correctly authorized users. FPR is the fraction of incorrectly authorized users, and complexity is the number of lines when the policy is written in a programming language. Unicorn archives TPR of about 90, 60, 90, and 70% for RBAC, ABAC, XACML, and STARBAC, respectively. The reported FPR was less than 1% for all the policy languages. In terms of complexity, Unicorn has generated substantially less complex policies only in the case of ABAC.

3.2 Network Security Policies

While the access control policies discussed in the previous section typically involve conceptual entities (e.g., users, resources, and operations), network policies are more specific to the underlying network topology and organization. Network security policies are expressed in terms of rules that involve network-level intricacies, such as host network identifiers and how the endpoints communicate with each other.

Example

Table 3.2 lists a set of network security policies. The first rule allows HTTP traffic originating from host 192.168.1.2 at port 3035 destined to host 192.168.1.3 at port 80. The second rule denies any TCP traffic originating from host 192.168.1.3 to host 172.50.50.23 at port 4444. Finally, the third rule allows the host 192.168.5.5 to communicate with any party on the network. Clearly, the first is the least privilege rule, as it only allows a specific type of traffic to flow between certain application ports of the endpoints. The second and third rules are less restrictive, with the latter being the least restrictive of the three.

ML approaches have been proposed to learn network access control policies. Typically, the training would be carried out on historical network access requests or communication flows.

Table 3.2 An example of a set of network security policies

Source IP address	Source port	Destination IP address	Destination port	Protocol	Access
192.168.1.2	3035	192.168.1.3	80	HTTP	Allow
192.168.1.3	–	172.50.50.23	4444	TCP	Deny
192.168.5.5	–	–	–	–	Allow

3.2.1   Firewall Rule Miners

One of the most critical elements in network security is firewalls. Firewalls are typically placed on the network boundaries to control and filter inbound and outbound traffic. Of course, networks could also be architected with micro-segmentation [133], in which case a firewall device guards the micro-segment. A firewall requires the network administrator to define the policy in terms of network rules. Typically, those include the classic 5-tuple, that is, source and destination IP addresses, the source and destination ports, and the transport layer protocol. However, the policy could incorporate more fields depending on the firewall's capabilities.

Authoring the firewall policy, however, is a daunting, time-consuming, and highly error-prone task. The slightest error in the policy rules could result in a cascade of network communication failures or even allow potentially malicious traffic to cross into the network. Therefore, approaches have been proposed to generate the rules for firewalls cost-effectively by providing high-level network requirements or analyzing the network traffic. In this section, we focus on the latter.

Golnabi et al. [69] proposed approaches to mine firewall rules from traffic logs. Their approach has three objectives: (1) mine the firewall rules from traffic logs using frequency analysis, (2) perform rule generalization to reduce the number of generated rules, and (3) identify decaying and dominant rules as observed in the traffic logs. The mining process has the following steps:

(1) Analysis of firewall rules: In this step, the algorithm takes as input the ruleset of a firewall.

Example

Here is an example of a policy rule specified in a Linux firewall.

```
-A INPUT -s 129.168.47.35 -p tcp --dport 80 -j DROP
```

The above rule specifies that any TCP packet originating from the IP address 129.168.47.35 to the local destination port 80 should be dropped. Here is an explanation of the rule components: -A: Append the rule to the ruleset. INPUT: for packets destined to local sockets. -s: source address. -p: protocol. –dport: destination port. -j: jump to target (in this case, to drop the packet).

From each firewall rule, the algorithm extracts the following fields: the protocol (i.e., TCP and UDP in the paper), the traffic direction (i.e., inbound or outbound), the source and destination ports, the source and destination IP addresses, and the target action (i.e., allow or deny). If any of the attributes mentioned above are absent from a rule, it is assumed that these attributes do not impact the rule.

(2) Firewall traffic log analysis: In this step, the algorithm processes the packet traffic log using ARuM and mining firewall log using frequency (MLF). MLF extracts the attribute features from each log entry and computes the number of packets that match every rule.

(3) Filtering-Rule Generalization (FRG): This step aims to minimize the number of the generated rules for efficiency purposes by aggregating them together. Rule aggregation is achieved through a decision tree method. The tree's root is the action (either allow or deny), and then follows the protocol (TCP or UDP in this paper), the traffic direction, the destination port, the source port, the source IP address, and the destination IP address.

(4) Ruleset generation: The fourth step combines the firewall ruleset (step 1) with the rules generated by applying MLF on the traffic log (step 2). The combination is carried out using an anomaly detection and resolution algorithm. The algorithm considers four anomaly categories:

- Rule Shadowing: A previous rule in the ruleset matches all packets of this rule; hence this rule is never activated.
- Rule Correlation: Two rules are correlated if they have different target actions, and the first rule matches some packets matching the second rule and vice versa.
- Rule Generalization: This is the case where a subsequent rule is a generalized version of the previous one, but they have different target actions.
- Rule Redundancy: A rule is redundant if its removal does not affect the overall policy. In other words, the rule is never activated.

(5) Dominant and Decaying Rule Detection: Then the probability distribution is computed for each generated rule using the firewall traffic log. In particular, the probability distribution of the rule is computed as $P = f/N$ where f is the frequency of the packet occurrence matching the rule, N is the total number of packets, and P is the probability distribution.

A similar approach was proposed by Tongaonkar et al. [229], which aimed to infer the high-level policy from low-level packet filtering rules and convey the analysis to the end-user. The approach has two steps:

(1) Priority Elimination: In this step, the algorithm generates a flattened ruleset from the provided iptables firewall ruleset. A flattened ruleset is a ruleset that captures the properties from the underlying low-level rules (such as ports and addresses), and the rules have no priority. If the rules in a ruleset do not have any priority, then any ordering of the rules does not change the overall policy. Thus, the rules are not overlapping (only one rule can match a given packet).

A naive approach to generating a flattened ruleset could result in an exponential number of flattened rules. To prevent such an exponential rule generation, a technique, called packet classification automaton based on directed acyclic graphs [230], is used. The idea is that a network node in the packet header is represented as a node in the graph (e.g., destination host). The leaf nodes correspond to target actions (i.e., allow and deny). The outgoing edges from each node are labeled with the values that the edge's source node can take. In addition, a node has an "else" edge that denotes the next state in the automaton. This edge is applied to a packet if no other outgoing edges (i.e., values) meet the property value of the packet under test.

Once the automaton is constructed, for every packet, there is a path from the root to a leaf of the automaton that determines whether the packet should be allowed or denied. In addition, every path is unique; thus, all the paths (i.e., generated rules) are not conflicting. **(2) Policy Inference**: The priority elimination step could generate a massive number of rules as the automaton could become overly complex depending on the scenario. This step first defines a complexity metric between two representations of the same flattened ruleset and aims to reduce the complexity of the generated flattened rules.

In order to reduce the complexity of the rules, one must merge them so to minimize complexity. Thus, the problem is to generate an equivalent representation of the ruleset with minimum complexity. This problem is equivalent to the minimum set cover problem, which is NP-complete. However, exploiting the properties of the flattened rule structure makes the problem tractable. We refer the interested reader to [229] for more details.

3.2.2   ML-Based Firewall Systems

Ertam et al. [68] proposed the idea of training an ML model that enforces the firewall policy instead of mining the ruleset. In particular, they conducted an experiment in which they trained a multiclass SVM using firewall log data. The firewall log data are composed of packet header features such as involved ports, bytes sent and received, and elapsed time. The target (i.e., the labels) is the action field, and it has four classes: allow, deny, drop, and reset-both. The difference between deny and drop is that in the former case, a message will be sent back to the sender (typically a "Destination Unreachable" ICMP message) letting it know that the packet is not delivered, whereas in the latter case, the firewall will silently drop the request without sending any notification message. Ertam et al. have experimented with SVM Linear, SVM Polynomial, SVM RBF, and SVM Sigmoid, achieving F1 scores of 75.4, 53.6, 76.4, and 74.8%, respectively.

Aljabri et al. [12] have conducted a similar study, experimenting with a broader range of ML models: K-Nearest Neighbors, Naive Bayes, J48, Random Forest (RF), and Neural Networks. At the same time, they performed feature correlation, and they found the most crucial feature that highly influences the model classification. They found that the RF classifier produced the highest accuracy of around 99.11%.

3.2.3   Network Security Policies for Traditional Networks

Drozdov et al. [70] proposed OLAPH, a system for online learning of anomaly detection (AD) policies from historical data. Their approach is based on *symbolic learning*. Symbolic learning refers to the learning of high-level knowledge, such as knowledge expressed as rules and logic programs like answer set programs [140, 141]. Such rules and programs thus use symbols, such as variables, and express relevant semantic relationships among different entities. Two advantages of symbolic learning are that (1) it requires less training data than other ML techniques; and (2) it is explainable as the generated expressions are in human-readable form.

OLAPH uses FastLAS [140] for learning the policies. FastLAS is a system for learning answer set logic programs, a special form of logic programs. FastLAS thus returns a set of logical rules. An example of such a rule is *allow* $\leftarrow$ *source_address*(192.168.5.5), *source_port*(5050), which implies that all traffic originating from 192.168.5.5 at port 5050 should be allowed. Of course, rule expressions can grow arbitrarily large, in that they may include large number of predicates; thus, the attribute search space could be huge. In FastLAS, one has to define the language bias, which defines which attributes may be used in the head of the rules and which ones in the body of the rules. We refer the readers to [140] for more details.

The input data are access requests, each of which may contain an arbitrary number of attributes. The user must provide a hyperparameter that sets the maximum number of features that FastLAS should select. Once the data are fed into FastLAS, the system applies a feature selection process where it identifies the most important features that significantly impact the decision (allow or deny). This process ranks features by *kurtosis*, which measures the tailedness of a feature's distribution.

The next step is to generate the learning task for FastLAS. FastLAS has to search the entire space of all the possible attribute combinations to learn the positive rules. The more the attributes, the larger the search space is. Thus, it is a typical combinatorial problem, which could require a very long time to return the results. OLAPH addresses this challenge by applying a domain-specific scoring function that the user must provide. Then the problem turns into an optimization problem, as FastLAS tries to find a solution (i.e., a set of symbolic rules) that minimizes the score.

The final component of OLAPH is online learning, where the system implements a procedure for updating the policy over time. Policy updates are an essential part of the process as the underlying requirements may change for several reasons, such as application updates or device replacements. OLAPH implements this mechanism by computing the distance between incremental windows (i.e., batches) of requests and the training set used to learn the current policy. This distance is called confidence level and is used to compute the *relearn indicator*. OLAPH uses the relearn indicator to determine when the policy should be relearned. Each rule is converted to a numerical vector using one-hot encoding, and the distance is the minimal value of the Manhattan distances between the one-hot vector and each of the one-hot vectors of the rules in the current training set. Then the indicator is computed as the mean of the maximum distances across the windows.

The relearn algorithm extracts two threshold values from the standard deviation of the learn indicator, which forms a range. If the relearn indicator falls out of that range, then the relearn process is initiated based on the new windows. FastLAS generates the new rules and sends them to a user administrator for review.

The experimental results show that OLAPH achieves a 0.8 AUC score. We refer the interested reader to the experimental section of [70] for more details on the experiments.

3.2.4 Network Security Policies for IoT

Over the years, IoT networks have become increasingly popular in many different environments, such as smart buildings, smart cities, smart homes, and even smart grids. IoT devices (smart locks, cameras, lighting, and sensors) are highly heterogeneous and from different vendors with proprietary control platforms (such as Philips Hue and SmartThings). Such limitations led to interoperability issues in operations across different platforms (i.e., domains), including access control. The second challenge for the security policy learning task is that IoT users may be reluctant to share their datasets (e.g., access logs) or choose to share part of the data. Training on incomplete training data results in problems with policy correctness and completeness of the learned policy.

Li et al. [148] have proposed an approach to address the two aforementioned challenges. They proposed FCAD, a federated learning-based cross-domain access decision method. FCAD eliminates the need for access control logs exchange between the domains. In particular, domains train their ML models using local data, and then a global model is developed by obtaining the calculated gradients from the local model. In other words, domains share only the computing gradient values, not local access control data.

In the first step, each participant (i.e., domain) generates its own model and then shares the model parameters (i.e., model weights) with the central server. FCAD uses a sequential model composed of four interconnected neural network structures where the first three use ReLU and the last uses Sigmoid activation functions. The last layer predicts the access request result (i.e., allow/deny). The authors have used the binary cross-entropy as the loss function. The training set used is the DS2OS (dataset #6 in Appendix).

The central server uses the model parameters received from each participant and updates the global model using the gradient aggregation scheme. This scheme measures the contribution of each participant. The higher the contribution, the more influence the participant has on the training of the federated model. The training of the federated model continues until the proposed loss function has a value that is less than a user-specified hyperparameter θ. The experimental results show that FCAD achieves a prediction accuracy of 83.6%.

3.3   Privacy Policy Contradiction Identification

Andow et al. [16] propose PolicyLint, a privacy policy analysis tool that identifies contradictions in privacy policies by simultaneously considering negation and varying semantic
levels of data objects and entities. PolicyLint automatically generates ontologies from a large
corpus of privacy policies and uses sentence-level NLP to capture both positive and negative
data collection and sharing statements. Figure 3.2 shows the steps followed by PolicyLint.

The first step is ontology generation, for which PolicyLint uses a semi-automated and
data-driven technique. It breaks ontology generation into three main parts. First, PolicyLint
performs domain adaptation of an existing model of named entity recognition (NER). NER
is used to label data objects and entities within sentences, capturing not only terms but also
the surrounding context in the sentence. Second, PolicyLint learns subsumptive relationships
for labeled data objects and entities by using a set of lexicosyntactic patterns with enforced
named entity label constraints. Last, PolicyLint takes a set of seed words as input and
generates data-object/entity ontologies using the subsumptive relationships discovered in
the prior step.

The second step is the policy statement extraction, which generates a concise representation of a policy statement to allow for automated reasoning over this statement. PolicyLint
represents data sharing and collection statements as a tuple (actor, action, data object, and
entity) where the actor performs some action on the data object, and the entity represents the
entity receiving the data object. PolicyLint extracts complete policy statements from privacy
policy text by using patterns of the grammatical structures between data objects, entities,
and verbs.

The last step for PolicyLint is to use ontology and extracted policies to detect contradictions. The first step is to simplify the policies by capturing shared information as a collection.
For the second step, privacy policies are modeled as a set of simplified policy statements.
A contradiction occurs if two policy statements suggest that entities both may and may not
collect or share a data object. Contradictions have two primary impacts when privacy policies are analyzed. First, the contradictions may result in incorrect automatically generated

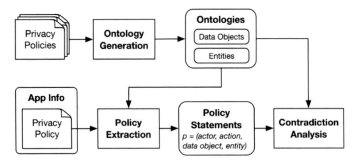

Fig. 3.2 Overview of PolicyLint [16]

decisions. Second, contradictions may impact a human analyst's understanding of a privacy policy, such as containing misleading statements.

PolicyLint is a really interesting approach for finding contradictions in policies. PolicyLint is built on top of current NLP techniques for NER and parsers. Since the paper's publication by Andow et al., the NLP community has made significant progress in tackling these tasks by introducing transformers and large language models. Maybe PolicyLint would perform much better if it were updated with state-of-the-art NLP techniques. Furthermore, as PolicyLint has false positives, this issue can also be addressed by using state-of-the-art NLP models. This might be interesting future work.

3.4   Adaptive Security Policy Learning Systems

Policy adaptation is a critical requirement in the policy learning pipeline. Policies evolve over time due to changes in the underlying environment. Relevant use case scenarios include changes in user or resource behavioral patterns (e.g., changes in the number of accesses, access request types, and generated network traffic) due to newly introduced users and resources or changes in role assignments. Another consideration is the availability of more training features (i.e., attributes), such as location information which may substantially improve the completeness and the correctness of the learned policy. Therefore, the training of the security policy learning model cannot be a one-time operation, but it must be a continuous process.

Karimi et al. [130] proposed ABAC-RL, an adaptive access control model based on Reinforcement Learning (RL) for IoT. RL does not require much pre-labeled data compared to supervised learning techniques. Also, the deployed agent learns from the environment in which it is deployed and can adjust to dynamically changing conditions.

ABAC-RL has two components: the Authorization Engine (AE) and the environment. The AE includes an agent RL agent that automatically adapts the ABAC policy using a feedback control loop while interacting with the environment. The environment comprises the users interacting with the IoT devices and the system administrators. Initially, the agent may have little or no knowledge regarding existing authorization policies. Then the agent learns the policies using a reward/penalty scheme from its environmental actions.

When the AE receives an access request, it checks the system's state and decides the action (i.e., permit/deny) based on the learned policy. Each state in the system is composed of four attribute tuples: attributes of the user requesting access, attributes of the object, environmental attributes, and the operation to be executed on the object. Once the AE has executed the action, the agent receives a reward or penalty based on the feedback provided by the users and adapts the learned policy accordingly. The agent is rewarded if the users agree with the authorization decision or penalized in the other case. The reward is calculated using the predefined reward function and affects the agent's future decisions. The authors proposed the following reward function (r_t):

$$r_t = \sum_{w \in owner(o_t)} \lambda_{TP} \cdot TP_w + \lambda_{TN} \cdot TN_w - \lambda_{FP} \cdot FP_w - \lambda_{FN} \cdot FN_w \qquad (3.2)$$

where

- r_t is the computed reward at step t.
- $owner(o_t)$ returns all the owners/admins of the object on which the user requests access at time step t.
- $\lambda_{TP}, \lambda_{TN}, \lambda_{FP}, \lambda_{FN}$ are constant weights.
- TP_w is 1 if the agent correctly permitted the access; 0 otherwise.
- FP_w is 1 if the agent incorrectly permitted access; 0 otherwise.
- FN_w is 1 if the agent incorrectly denied access; 0 otherwise.
- FN_w is 1 if the agent correctly denied access; 0 otherwise.

Defining a good reward function is extremely important for effective agent training. However, the agent will become as good as the training access requests it receives. If the set of unseen states (i.e., examples) is large, then the agent is expected to perform poorly on future requests. The authors have proposed an approach to reduce the space of unseen states. The idea is that the states are not necessarily independent, and there could be some that are actually related. The admins need to predefine hierarchical attribute relations used in identifying related states. Informally, two states are related if all their corresponding attribute values except one are the same. For instance, suppose that two access requests were made where all attributes are the same except the "age_group". The access was granted for the first request where the age_group of the requester is a teenager. For the second, the requester is an adult. If the admin has defined a relationship between the value "adult" and "teenager" where the "adult" is an ancestor of "teenager", then the access should be granted as well.

Argento et al. [18] proposed ML-AC, which uses ML techniques to adapt the access control policy based on users' behavioral patterns. For example, suppose the following policy is enforced:

```
<Rule 3: userA (ID: risk_analyst),
        resource (ID: RecordsDatabase-1),
    read, permit>
```

The rule states that $userA$ can read data from the records database. At the same time, $userA$ can perform an arbitrary amount of reading operations on the database. For several reasons, such operations could be anomalous (or suspicious); for example, $userA$ is fetching too much data or performing this action outside business hours. ML-AC aims to augment the policy by learning contextual features such as the number of accesses and the amount of accessed data.

ML-AC builds a behavioral model for each user, which consists of attributes of the access requests (i.e., user, resource/object, and action) as well as contextual information available to the access control system (e.g., location, time of day, and bytes sent). Behaviors are grouped into classes of interactions where each class has instances of normal user behavior. Anything outside those classes is considered abnormal. Random Forest (RF) [35] classifiers are used for each interaction class to predict whether a new instance (i.e., access request) belongs to a user class of interaction. In particular, each RF is an ensemble of Decision Trees (DT), each modeling the conditions that identify normal behaviors. Each DT outputs rules in the form *antecedent* $\Rightarrow$ *consequent* as shown in [183]. The antecedent expresses the rules in the form of logical conjunctions, while the consequent is either normal or abnormal.

Example

The following is an example of the DT output:

```
<Rule 4: userA (ID: risk_analyst),
   resource (ID: RecordsDatabase-1), read, #read<20, #byte<1 GB
permit>
```

The augmented rule has two additional attributes, $\#read$ and $\#byte$, based on the behavioral observations made by the DT. The access will be permitted if the user makes up to 20 requests fetching at most 1 GB of data within a specific time frame.

Once the behavioral profile of users is constructed, the next step is to monitor compliance with the policies. Policy monitoring must be performed for two crucial reasons. First, to detect any anomalous behaviors by users and react by quarantining users. Quarantine actions could limit a user's ability to access some resource or deny any action until the administrator's intervention. Second, policies naturally evolve over time due to several factors, such as promotions of employees that would thus need access to other types of resources and perhaps not require access to resources accessed before the promotion. For those reasons, an ML-based access control system must continuously look for changing behaviors.

Argento et al. [18] use the Ollinda clustering-based approach [216] to monitor policy compliance and update the policy. They use the k-means algorithm to formulate behavioral clusters and detect new ones.

3.5   Research Directions

Most approaches proposed for policy learning typically consider only two types of actions, namely allow or deny. However, the action set could be more extensive. For example, when suspicious access requests are identified, instead of completely denying access to the network, a policy could allow access but route the traffic through a suitable network security function (NSF) such as an intrusion detection system (IDS). Then, access decisions could be based on taking into account the feedback from the IDS.

Such a scenario leads to the following research directions:

- **Dynamic NSF orchestration policies**: We need an approach to decide when a NSF or chain of NSFs should mediate a particular traffic flow. In other words, deploy the NSF only when needed. ML techniques could be instrumental in solving this task. One could train an ML classifier on an IDS dataset and then run the observed flow through the classifier to see whether it is anomalous. In such a case, a dynamically generated policy must be enforced to route this flow through the IDS. A benefit of such an approach is minimizing the security overhead introduced by the NSFs and improving network performance.
- **Policy engine and enforcement**: The output obtained from NSFs results in valuable network awareness data, for instance, when a firewall has identified suspicious traffic. Such data should be sent to the policy engine to generate dynamic policies (e.g., route traffic through IDS, or drop the communication). In addition, features extracted from anomalous traffic flows could be used to enhance the learning process and thus improve the efficacy of the employed ML techniques and evolve over time. A challenge here is to see how to utilize the knowledge extracted from the different NSFs along with any training datasets and develop a suitable ML architecture.
- **Integration with Software-Defined Networks (SDNs)**: The foundation laid out by SDNs makes them a suitable environment for the policy engine. The policy engine could be implemented as a network application running on an SDN controller. Then one would need to design and implement interfaces (or protocols) to facilitate the communication of the NSF with the controller (e.g., see [77]). The policy engine could utilize such communication interfaces to configure the NSFs and collect their output automatically. The policy engine module needs to take action by generating appropriate policies and enforcing them in the data plane. An example policy could be to route the communication of endpoints through a NSF. Then the controller should collect NSF results and decide how to handle the flow.

- **Attack management policies**: Once an attack has been identified, a key decision is related to the steps to contain/stop the attack. Assuming a pool of potential actions, deciding which action or set of actions to perform highly depends on desirable objectives. For example, denying all communications is less viable when critical infrastructure services are involved. Once again, ML approaches could make a step toward solving such a problem. For instance, RL techniques could be trained on several attack scenarios and, given network awareness information, could decide which actions to take.

Software Security Analysis

4

Researchers in academia and industry have proposed numerous approaches to automatically analyze software to detect bugs and vulnerabilities. Approaches can be divided into two main categories: static analysis and dynamic analysis. Static analysis aims at analyzing software without executing its code. It is usually performed by dissecting the different resources of a binary file or source code and analyzing each component in detail to expose bugs and vulnerabilities. Dynamic analysis refers to techniques that evaluate software correctness by dynamically executing the target software. Such techniques are performed by observing the program's behavior while running and monitoring its activities and processes. Dynamic analysis objective is to spot anomalous events that can expose the existence of vulnerabilities and unintended software behaviors. However, it is important to note that to ensure that software is secure, also the specifications need to be analyzed to identify mistakes, such as inconsistency and underspecifications.

In the chapter, we first present approaches that use ML techniques for static analysis. We then focus on fuzzing techniques, which are widely used for dynamic analysis. Because of its popularity, many approaches have been proposed to enhance fuzzing with ML techniques—many of which focus on the problem of fuzzing coverage. We then discuss approaches using natural language processing (NLP) techniques for the analysis of software specifications and other software documentation written in natural language. This is an important emerging area as specifications often consist of large-scale documents written in natural language, as is the case for example of the 5G cellular network standard specifications. NLP techniques allow one to quickly identify issues in the specifications. However, NLP-based analysis of specifications is also useful for other tasks, such as identifying risks in protocols and attack generation, which we also discuss in this chapter.

© The Author(s), under exclusive license to Springer Nature Switzerland AG 2023
E. Bertino et al., *Machine Learning Techniques for Cybersecurity*, Synthesis Lectures on Information Security, Privacy, and Trust,
https://doi.org/10.1007/978-3-031-28259-1_4

4.1   Static Analysis

Static analysis refers to techniques and tools designed to extract facts from a "program's source code, without executing the program in question" [227]. In the context of software security, facts extracted by using static analysis include misuse of cryptographic APIs, taint analysis, which identifies and monitors the use of values that could be tainted with malicious data, and bounds checking, which ensures that accesses to arrays or memory locations fall within the bounds that the programmer intended. Because of its relevance, the area of static analysis has been extensively investigated over the years, and many tools are today available. We refer the reader to [228] for an interesting historical perspective and an outlook toward the future, and to [74] for a detailed comparison of industrial static analysis tools.

In what follows, we first summarize the results of a recent survey that covers the application of ML techniques to source code analysis for different tasks, including the identification of vulnerabilities. We then summarize three recent approaches which use ML in a novel way compared to other approaches.

4.1.1   A Survey on Machine Learning Techniques for Source Code Analysis

Sharma et al. [209] recently published a comprehensive survey on ML techniques for source code analysis. The survey reviews a corpus of 479 papers in the existing literature on ML techniques applied to 12 different source code analysis categories: code completion, code representation, code review, code search, dataset mining, general, program comprehension, program synthesis, quality assessment, refactoring, testing, and vulnerability analysis. In the following, we briefly overview the approaches for some of the source code analysis categories following the survey structure. We primarily focus on the categories more relevant from a software security standpoint, trying to provide some insights into the high-level ideas. Most approaches follow the typical ML pipeline: dataset collection, feature extraction, and model training.

Code representation. One cannot directly give source code as input to ML algorithms as they are mainly designed to work with numerical representations. Thus, different techniques have been proposed to address this challenge. Such techniques typically convert the source code into an intermediate representation, apply a tokenization mechanism, and then produce a vectorial representation. Examples of intermediate representations are Abstract Syntax Tree (AST), Control Flow Graph (CFG), Inter-procedural Value Flow Graph (IVFG), and LLVM-IR (LLVM Intermediate Representation). In some cases, the tokenization is directly applied to the source code. Once the source code is translated into an ML-compatible representation, the approaches in this category mainly employ DL techniques, such as LSTM and GRU, DNN, and CNN, to learn a model of the source code. Some approaches also use combinations of multiple models.

Testing. ML techniques for source code testing aim to investigate the target code to detect functional or non-functional bugs. Most approaches in this category leverage existing datasets, such as the PROMISE [182] dataset. Some approaches build approach-specific datasets from publicly available GitHub repositories. Later in this section, we also present in more detail a unique approach by Ahmadi et al. [3] that, instead of using external datasets, leverages the target code itself as a dataset to train the ML model. The extracted features typically represent meaningful code metrics, such as the number of lines of code or of function calls. Several approaches use techniques to reduce the feature space, such as PCA and Sequential Forward Search (SFS). Often, this step also involves translating the code into an ML-compatible code representation, as discussed above. Finally, the approaches train ML models to detect buggy code. The most popular techniques here include traditional ML models, such as Decision Tree, Random Forest, Support Vector Machine (SVM), and AdaBoost, as well as DL models, such as RNN, among which LSTM is the most common, and CNN.

Program synthesis. Among program synthesis techniques, the survey primarily focuses on *program repair*, which aims to patch erroneous code automatically. Approaches in this category often build their own dataset by mining existing software project revisions from open-source libraries and repositories (e.g., GitHub). Some approaches collect datasets of buggy code and manually introduce the fixes using known fixes corpus. All approaches aim to build a dataset containing pairs of buggy and corresponding fixed code components. For feature extraction, most approaches use similarity metrics to detect similar bug patterns based on word embeddings (e.g., Word2Vec) or structure-based representations (e.g., AST). Traditional ML techniques for program synthesis include Decision Tree, Random Forest, SVM, and Logistic Regression, whereas approaches that employ DL use RNN models, such as LSTM.

Quality assessment. ML techniques for quality assessment aim to investigate the target code to detect code smell and replication. Code smells are a result of poor or misguided programming; they typically stem from the failure to write the code in accordance with necessary standards. Datasets for quality assessment are typically generated by manually annotating existing collections of code smells and replication. Few approaches leverage existing tools to automatically detect code smells and label their dataset. For code smell detection, the most common feature selection processes focus on object-oriented metrics, such as the number of lines of code, number of methods, function parameter count, and depth of nested conditionals, whereas for code replication, the features are typically based on the textual properties of the source code (e.g., word embeddings). Traditional ML techniques (e.g., Decision Tree, Support Vector Machine, Random Forest, Naive Bayes, Logistic Regression, Linear Regression, Polynomial Regression, and Multi-layer Perceptron) and DL methods (e.g., ASN, DNN, CNN, and RNN) have been proposed for both code smell and code replication detection. Some approaches also experiment with ensemble methods, such as Majority Training Ensemble and Best Training Ensemble, for code smell detection.

Vulnerability analysis. ML models for vulnerability analysis aim to identify potential vulnerabilities in source code. Some of the approaches in this category use existing labeled datasets for PHP, Java, C, C++, and Android, whereas others generate their own datasets by collecting open-source applications and manually inserting vulnerabilities. The feature selection is rather diverse across the approaches and can be categorized into source code metrics (e.g., number of instructions and number of blank lines, and inheritance tree depth), data/control flow (e.g., CFG and AST), repository and file metrics (e.g., programming language, fork count, commits number), and code and text tokens (e.g., character count and diversity, `if`, `while`, and `for` count and complexity). As for the other categories, ML models for vulnerability analysis range from traditional ML models, including Naive Bayes, Decision Tree, Support Vector Machine, Linear Regression, Decision Tree, and Random Forest to DL models, such as CNN, ANN (specifically BP-ANN), and RNN (LSTM and Bi-LSTM).

4.1.2 Recent Approaches

An interesting ML-based approach for static analysis is by Ahmadi et al. [3]. The approach, referred to as FICS, leverages a two-step clustering model to detect code inconsistencies using the target source code itself (note that code inconsistency refers to code snippets that are semantically and logically equivalent but different in their implementation). FICS first converts the source code into an LLVM intermediate representation and then applies intraprocedural data flow analysis to extract code pieces in the form of Data Dependency Graphs (DDG). Such code pieces, referred to as *constructs*, represent self-contained subroutines included in a function. The nodes of the constructs represent the code instructions, whereas the edges express the hierarchy of the instructions. Subsequently, FICS abstracts the constructs to a more general representation by removing constants and literals to better support clustering. The first step of clustering aims to detect similar constructs. To do so, FICS employs an approach referred to as *bag-of-nodes*, namely the constructs are represented only using their nodes (i.e., instructions), thus ignoring the edges. Ahmadi et al. suggest using the connected-component algorithm as it achieves better performances with an unknown number of expected clusters. For this step, the cluster similarity threshold is between 0.8 and 0.95, as the goal is to cluster together similar but not necessarily equal constructs. For the second step of clustering, FICS reintroduces the edges in the constructs representation and employs Graph2Vec, a graph embedding mechanism, to translate the constructs into a vectorial representation. Then, FICS independently performs clustering within each cluster identified in the first step. For this step, the authors suggest using a very high similarity threshold ("1 or very close to 1") as they aim to separate constructs that are similar but not exactly the same. The output is a set of similar but different constructs that can potentially represent code inconsistencies. Finally, a manual analysis is required to discern false inconsistencies from true ones. The advantage of the approach by Ahmadi et al. is that it leverages the source code itself. Thus, unlike other approaches addressing the same problem, it does not require any dataset, is not limited to any specific type of inconsistencies, and can detect one-to-one inconsistencies. However, the approach requires a large codebase for the ML model to be effective, and incorrect code sections can only be detected if there is a correct version of the same sections in the code, thus generating an inconsistency.

Another example of a ML-based static analysis is represented by the tool GLACIATE [151], which has been designed to detect flaws related to the implementation of the basic login-password authentication in Java mobile applications. More specifically, GLACIATE detects vulnerabilities related to (i) insecure password transmission, that is, passwords that are not encrypted, or encrypted but not hashed before encryption; (ii) lack of server validation, that is, the server certificate is not properly validated (or not validated at all), or the host name on the certificate is not validated against the host portion of the server's URL; (iii) repeatable timestamps for timestamp-based authentication. Given only a small amount of training data, GLACIATE creates detection rules automatically. It generates enriched call graphs for the apps and groups similar enriched call graphs into different clusters, and mines the patterns of flaws in each cluster to obtain templates of insecure implementation. GLACIATE then uses a forward and backward program slicing to locate the code part of password authentication in an Android app, and compares it with the obtained templates to check whether the implementation is insecure.

An interesting novel approach that enhances the efficiency of static analysis with the use of ML is represented by the Goshawk tool [150]. Goshawk has been designed with the goal of enhancing the accuracy of static analysis in the presence of memory bugs due to custom memory allocation, while also reducing the time required to statically analyze very large code bases. To address such a goal, Goshawk uses the concept of structure-aware and object-centric memory operation synopsis (MOS). A MOS is a tuple that comprises a primary function name, a primary property (allocation or deallocation), and a list of dynamically managed memory objects occurring in either a return value or parameters of the function. It summarizes the structural relations of memory objects in memory management (MM) functions. Because how a memory object should be operated on by an MM function depends on its structure, a MOS focuses on the object (hence object-centric) and its structure (hence structure-aware). Moreover, a MOS not only captures the structure of memory objects but also describes the function action (allocation or deallocation) against each object. By using MOS, a bug detection tool does not need to explore the internal implementation of all MM functions but can still precisely model the dynamically managed memory objects.

A significant requirement for MOS-enhanced bug detection is, however, to first identify all MM functions, as these functions are the basis of MOS. To address such requirements despite the fact MM functions follow various implementation styles in different kinds of source code projects, Goshawk uses an accurate and yet very efficient identification technique that combines natural language processing (NLP) and data flow analysis. The Goshawk memory function identification technique is organized according to two main steps. The first step uses an NLP-assisted classification against function prototypes in source code to categorize functions as MM-relevant or MM-irrelevant. The insight here is that a function prototype is often human-readable, and the natural language semantics of the prototype usually reflects the functionality. Experiments have shown that by using this semantic information, Goshawk has been able to approximately classify MM functions in a short time even out of millions of functions. The second step applies a data flow analysis against the implementation of each MM-relevant function identified by the first step. The data flow analysis checks whether the function does indeed perform memory allocation/deallocation using known memory allocators/deallocators (which are previously defined manually). The combination of those two steps achieves both efficiency and accuracy. Efficiency is achieved because the NLP analysis prunes irrelevant functions (i.e., functions not related to MM), for which a detailed analysis is thus not required. Accuracy is achieved because the MM-relevant functions are analyzed in detail by using a static analysis.

The NLP-assisted classification starts from parsing the source code of the tested project to extract all function prototypes, and chooses the prototype with at least one pointer (either a return value or a function parameter). The extracted prototypes are then sent to a ULM-based segmentation process [138], and each is split into a list of subwords. Then, Goshawk converts the list of subwords into a numeric vector by utilizing a Siamese network [37] with a transformer encoder [235] trained with manually labeled MM and non-MM functions. Finally, Goshawk computes a similarity score between the numeric vector of a function and three reference vectors, which separately indicate memory allocation function, memory deallocation function, and non-MM function type. According to the similarity score, Goshawk classifies the tested functions as one of the above types. We refer the reader to [150] for details about the model training.

Experimental results show that Goshawk is highly effective. For example, Goshawk has been able to detect 40 new memory bugs in the Linux Kernel.

4.2   Fuzzing Techniques

Fuzzing is an approach that feeds randomly generated inputs to the target software and dynamically evaluates its behavior. New inputs are continuously produced as mutations of previously generated inputs and fed to the target software until the fuzzing loop is interrupted. Malformed inputs, namely inputs not compliant with the expected input syntactically or semantically, can trigger unexpected behavior and uncover potential vulnerabilities. For instance, one of the most common scenarios is a malformed input that causes a crash in the target software. A malicious entity could leverage this type of input to carry out DoS attacks. In what follows, we first give a high-level overview of the main steps involved in a fuzzing testing framework and then discuss how ML techniques have been used to improve each step (see Fig. 4.1). For detailed information about each step, we refer the reader to [152].

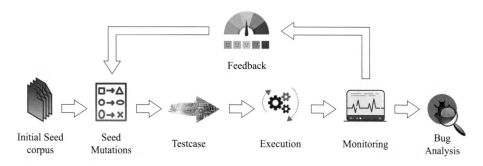

Fig. 4.1 Fuzzing main steps

4.2.1 Fuzzing Steps

4.2.1.1 Seed Generation/Selection

The first step of the fuzzing process consists of selecting a seed corpus from which to generate the inputs or testcases. Typically, the initial corpus can be randomly generated, obtained from a set of valid inputs to the program, or generated according to a given grammar. After the first selection, the fuzzing framework selects seeds from the set of previously generated testcases based on a feedback function that allows one to select testcases more likely to execute unexplored portions of the target software code. Once the seed corpus is selected, the fuzzing process moves to the second step, that is, seed mutation. Note that when selecting the new seeds, the fuzzing framework can also choose (typically randomly) not to mutate any previous testcases, but to generate entirely new seeds. This method allows one to avoid situations in which the fuzzing framework keeps mutating the same inputs without, however, never really moving forward in the exploration process.

4.2.1.2 Seed Mutation and Testcase Selection

In the second step, the fuzzing framework mutates the selected seeds to generate potentially malformed testcases. Several mutation strategies have been proposed to alter the structure and content of the selected seeds. Some of the most commonly used mutation strategies in state-of-the-art approaches, such as AFL [252] and its evolution AFL++ [79], change the seeds at the bit/byte level and include bit flips, byte flips, arithmetic mutation, block deletion, block swapping, etc. Once the testcases are generated, the fuzzing framework runs the target software with each testcase and evaluates the software execution.

4.2.1.3 Execution and Monitoring

In this step, the fuzzing framework executes the target software using each selected test-case as input. As the testcases are randomly generated, their syntax and semantics can be significantly diverse and potentially execute different sections of the target software. Some testcases might be quickly discarded, while others might trigger unexpected behaviors. To capture such unexpected instances, the fuzzing framework monitors the target software execution and signals anomalous events (for instance, through logging mechanisms). Execution information is then passed to the feedback function to select the corpus seeds for the subsequent execution. Monitoring for anomalous events is a critical task as it can expose existing bugs and detect potential software vulnerabilities.

4.2.1.4 Feedback (Fitness) Function

The feedback, or fitness function, is a key element in the fuzzing framework in that it allows one to evaluate the execution results of each testcase and detect the testcases that enable better exploration of the target software. One can opt for different feedback functions based on how much information about the software is available. One of the most popular techniques is code coverage, namely the amount of target code covered during the execution of a testcase. The intuition is that testcases that increase the code coverage are more likely to lead to the execution of additional code and, consequently, expose higher numbers of bugs and vulnerabilities. However, in some cases, it is not possible to have access to code coverage data, and it is necessary to employ techniques that infer execution information based on correlational measurements (e.g., execution time). According to the feedback function, the fuzzing framework selects the new corpus seeds, and the fuzzing loop starts again.

4.2.1.5 Bug Analysis

While fuzzing aims to expose bugs, it is also critical to analyze the exposed bugs and potentially understand their cause. Their understanding can help pinpoint the source of the bugs and also reason about potential exploitability. This could help to automatically evaluate if a bug is, in fact, exploitable and thus represents an actual vulnerability of the target software. Currently, this task is mostly performed manually, which entails potential errors and oversights.

4.2.2   ML-Based Fuzzing

We now discuss the ML techniques proposed for each step of the fuzzing process separately. Note that when applying ML techniques, the target domain is critical to the choice of the technique. For instance, when considering one step of the fuzzing process, the techniques used for document parser fuzzing might differ entirely from those used for fuzzing a network communication protocol.

4.2.2.1 Seed Generation/Selection

Generating and selecting high-quality corpus seeds can significantly improve the fuzzing performances when compared with merely randomly generated seeds. The reason is that the quality of the testcases produced from the seeds depends on the quality of the seeds and can avoid numerous fruitless fuzzing loop executions that do not explore any interesting software code sections. Thus, several ML-based approaches have been proposed to enhance seed generation and selection.

Reddy et al. [187] propose an approach to automatically generate program inputs that are both diverse and valid, leveraging an on-policy reinforcement learning (RL) algorithm. They assume the presence of an input generator specific to the target program input domain and its implementation to be accessible to introduce additional RL-related instructions. The generator constructs each input based on a sequence of choices taken at choice points. Thus, the intuition is to model the problem of generating diverse valid inputs as an RL problem as follows: the choices represent the typical actions in an RL problem, whereas the sequence of previous choices defines the state. Thus, the agent, referred to as the learner, aims to maximize the reward function by making the choices that lead to the best inputs, namely inputs with the highest potential to improve the feedback function (e.g., code coverage). Reddy et al. define the reward function as a combination of three weights associated with whether the generated input is unique (i.e., different from all previous ones), valid, or invalid. The first two weights are positive numbers to promote diverse and valid inputs, whereas the last is negative to demote fruitless inputs. As the fuzzing process runs, the learner learns how to guide the generator to construct the best possible inputs based on the feedback of previous inputs. Based on their insights, Reddy et al. implemented RLCheck, a framework to generate diverse valid inputs based on the on-policy Monte Carlo [220] RL approach. RLCheck outperforms similar existing approaches significantly, proving the tremendous impact RL can provide in generating effective inputs for fuzzing frameworks.

Wang et al. [237] propose Syzvegas—an ML-based extension of the kernel fuzzing approach Syzkaller [90]. They recognize two fundamental tasks in kernel fuzzing: (1) selecting the most rewarding fuzzing task between seed generation, seed mutation, and seed triage (evaluate and minimize the current seed to ensure that the next mutation can achieve at least the same code coverage); (2) selecting the most valuable seeds for mutation. While Syzkaller employs a fixed strategy, Wang et al. propose to model the problem as an adversarial multi-armed bandit (MAB) RL problem, where the different tasks represent the "arms" of the bandit. Such an approach allows Syzvegas to continuously learn as the fuzzing process proceeds and improve its decisions throughout time. The specific reward function is adjusted according to the executed task and is based on achieved code coverage and execution time. The objective is to maximize the former and minimize the latter. Wang et al. show that their ML fuzzing approach can significantly outperform previous state-of-the-art kernel fuzzing frameworks, such as Syzkaller and HFL [136].

Wang et al. [238] propose a similar idea to model the problem of selecting the best seed as a MAB RL problem. However, they also use hierarchical seed clustering that allows them to incrementally cluster seeds based on code coverage information. The reward for each seed is based on the code coverage achieved during its execution and its rareness, which expresses how rarely other seeds hit the code hit by the seed in consideration. The intuition is that seeds that hit rarely explored paths should be prioritized. Their fuzzing framework, referred to as AFL-Hier, is compared against existing AFL-based frameworks over the programs provided in the CGC benchmark and seems to perform significantly better than all its predecessors.

4.2.2.2 Seed Mutation and Testcase Selection

Mutating the seeds to produce testcases that are more likely to increase code coverage is fundamental to improving the fuzzing performance. A seed that previously increased the code coverage might result in testcases that are immediately discarded if the wrong section of the seed is mutated. Thus, ML techniques can help evaluate which section of the input to mutate (where) and what mutation strategies to employ (how) to produce better testcases.

Rajpal et al. [185] explore whether it is possible to augment the AFL [252] seed mutation strategy with ML learning techniques based on the previous mutation history of executed testcases and code coverage data. They propose ML models that intelligently select specific bytes to mutate in the selected seeds, leaving the other bytes unchanged. The goal is to increase code coverage minimizing the risk of mutating the seed portions that enabled higher code coverage in a previous iteration. They experimented with different models based on LSTM and sequence-to-sequence methodologies. These two approaches can effectively support long and varying-length inputs and handle the sequential nature of the input (a sequence of bytes potentially correlated). Before mutating a seed, the fuzzing framework queries the ML model to learn which seed bytes are useful and focus the mutation on the bytes deemed not useful. The objective is to maximize code coverage. The model must be trained on a dataset containing inputs and related code coverage information. To obtain such a dataset, an approach is to run an initial AFL campaign on the target software and then apply their augmented version of AFL. While the results are not overwhelming (only slight improvements are reported compared to the base version of AFL), the proposed ML-based approach improves the fuzzing mutation strategy. Given the results shown by Syzvegas and AFL-Hier for the seed selection task, one could also experiment with a MAB reinforcement learning model to dynamically enhance the fuzzing mutation strategy.

Similarly, She et al. [202] propose to use surrogate neural network models to learn approximations of real-world program branching behaviors through neural program smoothing and leverage such models to guide and optimize fuzzing mutation strategies. They implement their ideas in a fuzzing framework called NEUZZ. She et al. recognize that gradient-guided (e.g., gradient descent) optimization has significant potential to aid fuzzing approaches. However, they point out that it cannot be directly applied to fuzzing real-world programs as they often entail discontinuous branching behaviors. In other words, a real-world program often involves branching conditions on specific values with specific discrete output values. These branching behaviors translate into gradient functions with ridges and plateaus that

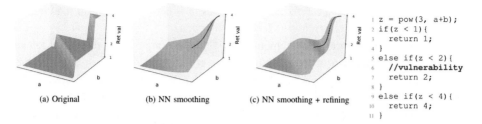

(a) Original   (b) NN smoothing   (c) NN smoothing + refining

```
1 z = pow(3, a+b);
2 if(z < 1){
3    return 1;
4 }
5 else if(z < 2){
6    //vulnerability
7    return 2;
8 }
9 else if(z < 4){
10    return 4;
11 }
```

Fig. 4.2 NEUZZ program smoothing methodology

gradient-based techniques cannot efficiently model. Therefore, She et al. propose to leverage feedforward neural networks (NN) to generate a smooth function of the program's behavior that can minimize the branching discontinuity. Given the complexity of the approach, we report a figure from the original paper to better clarify the authors' methodology. Figure 4.2 displays different functions representing the code snippet on the right. The original function image exposes the discontinuous behavior mentioned above, whereas the NN smoothing image shows the function profile after the smoothing process.

The smoothed function is then suitable for gradient-guided optimization. The authors highlight that NNs are advantageous as they can model complex non-linear program behaviors, support efficient computation of their gradients, and generalize and predict the program behavior with unknown inputs based on known similar inputs. For their experiments, the authors train the NN models using traces previously generated with existing fuzzing techniques, such as AFL. The traces contain program inputs and related behaviors expressed as exercised program control flow edges. A loss function based on distance metric is then used to train the NN model. Once trained, NEUZZ leverages the NN model to identify mutation locations in the fuzzing testcases that can maximize the number of program control flow edges exercised and, thus, potentially expose a larger number of bugs. As NEUZZ executes, the NN model also incrementally learns new program behaviors from new inputs. This neural program smoothing technique allows NEUZZ to outperform existing fuzzing approaches on both program datasets, such as LAVA-M and CGC, and real-world programs, such as *strip* and *nm*, also leading to 31 previously unknown bugs and two CVEs.

4.2.2.3 Feedback Function

Selecting the most effective corpus seeds for the next fuzzing iteration can increase the probability of finding software bugs. Standard code coverage has two main limitations: (1) it entails no understanding of what type of code is covered, which means that an increased code coverage might not lead to more exposed vulnerabilities—for instance, increased code coverage involving authentication routines might be more relevant from a security standpoint; (2) It is not always possible to collect code coverage information, such as in cyber-physical system black-box fuzzing. Therefore, in some cases, it is necessary to resort to alternative methods to collect information for the feedback function.

An interesting use of the ML technique to guide the fuzzing process through a domain-specific feedback function is proposed by Cheng et al. [44]. They target cyber-physical systems (CPS) by building an ML-guided actuator fuzzing. Their intuition is to use ML algorithms to model the behavior of the target CPS based on physical data logs that characterizes its normal behavior. Such an ML model is then used to determine which testcases are more likely to cause the system to reach an unsafe state. In this case, a testcase represents a possible configuration for the system, and the configurations are mutated to produce new testcases. The ML model of the CPS can predict the effects of the input configuration given a certain system state. At the end of each execution, the feedback function computes how close the predicted state is to an unsafe state. The predicted value is then used to select the best candidates for the subsequent execution. For this domain, Cheng et al. suggest using LSTM networks as these support the learning of longer term dependencies in the systems' input configurations and consequent state, and the support vector regression (SVR) [65] model to handle continuous values. The training dataset is a log dataset that contains information on the systems' normal behavior in a format that highlights the relationship between the systems' states and configurations, and the evolution of the states. The two models seem to perform very similarly in terms of the time necessary to reach the same unsafe states, so, for this domain, the authors could identify no specific advantage in using one model over the other. However, fuzzing based on either model significantly outperforms random fuzzing. Through their work, Cheng et al. show that ML techniques can be used to model the behavior of devices with limited capabilities based on previous execution logs. Although applying the same approach for devices with much more complex capabilities (e.g., smartphones) could be very challenging due to the model learning requirements, the authors' research could be extended to general IoT infrastructures and enable the fuzzing of IoT devices.

A novel ML-based approach to provide feedback has been recently proposed by Han et al. [95] in the context of LDGFuzzer, a fuzzing framework to detect unsafe combinations of control parameters for drones. Control systems of drones allow one to support a variety of missions and deal with different flight environments via configurable control parameters. However, such flexibility introduces vulnerabilities. One such vulnerability, referred to as range specification bugs, originates from the fact that even though each individual parameter receives a value in the recommended value range, certain combinations of parameter values may affect the drone's physical stability. To detect incorrect parameter configurations, LDGFuzzer applies metaheuristic search algorithms for mutating configurations. One issue, however, is how to effectively and efficiently obtain feedback from the tested configurations. Possible approaches can be based on realistic or simulated flight executions. However, because of the large number of control parameters, each with a wide value range, changing the parameter values to generate configurations and validating all these configurations is inefficient. Completing the entire validation procedure may then end up requiring hours or even days. To address such a challenge, LDGFuzzer uses a state generation approach that leverages a machine learning algorithm to train a state predictor, based on the LSTM technique, that based on the current position of the drone and a set of values for the control parameters predicts the next drone state. The predictor is trained with a set of flight logs collected from a simulation. However, this is cost incurred only once. Experimental results show that the total time required by LDGFuzzer, from generating the data to searching out 1,000 incorrect configurations, was only about 25 minutes, as the simulation does not have to be repeated for each configuration.

4.3   NLP-Based Techniques for Specification Analysis

NLP is a subfield of ML, mainly concerned with preprocessing and understanding human language by machines. With the improvement of state-of-the-art NLP techniques by large language models (LMs), such as BERT [63], and TL, a few approaches have been recently proposed using specialized NLP tasks for security. Furthermore, as software specifications, developer's notes, and policy guidelines often include large amounts of unstructured text that have to be manually analyzed, NLP techniques are now being investigated in order to automatically carry out such analyses. In what follows, we discuss a few use cases of NLP techniques for security. These cases range from the very rudimentary task of detecting some privacy-related settings to the task of generating 4G LTE testcases and to the difficult task of extracting formal models from the specifications.

4.3.1   Finite State Machine Extraction

Automated attack discovery techniques, such as attacker synthesis or model-based fuzzing, provide powerful ways to ensure network protocols operate correctly and securely. Such techniques, in general, require a formal representation of the protocol, often in the form of a finite state machine (FSM). Unfortunately, many protocols are only described in English prose, and implementing even a simple network protocol as an FSM is time-consuming and prone to subtle logical errors. To address such an issue, Pacheco et al. [175] have developed a data-driven approach for extracting FSMs from RFC documents. The approach is a hybrid with three steps: (i) large-scale world representation learning for technical language, (ii) zero-shot learning for mapping protocol text to a protocol-independent information language, and (iii) rule-based mapping from protocol-independent information to a specific protocol FSM (overview shown in Fig. 4.3). They apply this approach to six different protocols and use the FSMs to synthesize attackers as a representative technique for protocol security.

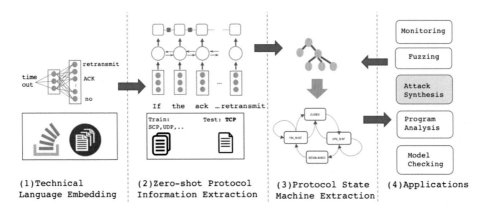

Fig. 4.3 Overview of FSM extraction from documents [175]

4.3.1.1 Technical Language Embedding

The first step for the NLP-based FSM extraction method is to learn the technical embedding of the underlying language. In recent years, in the NLP community, pre-trained language models (LM) provide the best way to derive contextualized representations of text, while allowing the users to fine-tune these representations for any downstream task. The most prominent example of such models is BERT [63]. BERT is built using a transformer, a neural architecture that learns contextual relations between worlds in a word sequence. A transformer network includes two mechanisms, an encoder that reads the input sequence, and a decoder that predicts an output sequence. Unlike directional models that read the input sequentially, transformer encoders read the whole sequence at once. To learn the representations, BERT uses two learning strategies, masked language modeling and next sentence prediction. In their approach, Pacheco et al. use the full set of RFC documents that are publicly available to pre-train BERT. These documents cover different aspects of computer networking, including protocols, procedures, programs, concepts, meeting notes, and options. The resulting dataset consists of 8,858 documents and approximately 475M words.

4.3.2   Zero-Shot Protocol Information Extraction

For the zero-shot protocol information extractions, Pacheco et al. aim to design a system that can adapt to new, unobserved protocols without re-training the system. To achieve this and parse the documents, they use a sequence-to-sequence model that receives text blocks as input, and outputs a sequence of tags corresponding to a simple grammar. For each textual unit in the input, a set of features such as vocabulary, capitalization patterns, logical and mathematical expression patterns, and position features are learned. They consider two models to learn the sequence-to-sequence mapping: a linear model referred to as LINEARCRF, and a neural model based on BERT embedding which is called NEURALCRF. The output for this step is an intermediate representation.

4.3.2.1 FSM Extraction

The intermediary representation obtained using the LINEARCRF or NEURALCRF model is not an FSM. Therefore, Pacheco et al. design a simple algorithm that extracts states S and transitions T by scanning the intermediary representation. The algorithm is based on several heuristics, and we refer the reader to the original paper [175] for details on these heuristics.

4.3.2.2 Attacker Synthesis

For showing a use case for the extracted FSM, they show the attacker synthesis. LTL attacker synthesis is a popular formal method problem, where a vulnerable part of a program or a system is determined based on a property, typically expressed in linear temporal logic (LTL). Suppose that $P \parallel Q$ is a system consisting of some programs P and Q, and ϕ is an LTL correctness property which is made true by the system. In the threat model where Q is the vulnerable party of the system, the attacker synthesis problem is to replace Q with some new *attacker* having the same inputs and outputs as Q, such that the augmented system behaves incorrectly and the system violates ϕ. One of the most interesting results of this approach is that the extracted FSM is not complete and the authors could not extract the complete FSM of the protocols. For example, in the case of DCCP from the complete FSM which has 34 transitions, this framework can extract a maximum of 24 transitions, among which 15 are correct, 1 is partially correct and 8 is incorrect. To use these partial FSM, the authors needed to modify an existing attacker synthesis tool named KORG.

4.3.2.3 Limitations and Discussion

One important result of this work is that it shows that a completely automated approach may not be enough to extract a complete FSM. The reason is that these complete FSM or canonical FSMs are created based not only on RFCs but also on input from experts with exposure to protocol implementations, and often also rely on analyzing the code. RFCs contain ambiguities, and unspecified behaviors that human experts solve in creating the canonical FSM. Moving forward, there are two approaches that can be taken (i) either design techniques that make it possible to create specs that are complete, or (ii) design NLP approaches that take human knowledge into account as well. From our perspective, applying a more human-in-the-loop approach is the viable path in this regard.

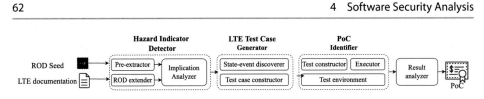

Fig. 4.4 Overview of ATOMIC (redrawn from Fig. 3 in [46])

4.3.3   4G LTE Testcase Generation

Chen et al. [46] propose an approach to use the technical documentation of cellular network protocols to generate testcases. They use the concept of *hazard indicators* (HIs)—a statement that describes a risky operation (e.g., abort and ongoing procedure) when a certain event happens at a state, which can guide a test on the system to find out whether the operation can indeed be triggered by an unauthorized party to cause harm to the cellular core or legitimate users' equipment. Based on this idea, they propose an approach, referred to as Atomic, that utilizes NLP and ML techniques to scan a large amount of LTE documentation for HIs. The HIs discovered are further parsed and analyzed to recover state and event information for generating testcases. These testcases are utilized to automatically construct tests in an LTE simulation environment. Figure 4.4 shows an overview of the approach.

4.3.3.1 Atomic

The approach is based on the observation that some risky operations, as described in the cellular network documentation, can be triggered in an insecure way, so identification of these operations and their conditions can help determine whether they have been properly protected and whether a vulnerability is present and can be exploited. The authors show this process can be automated with a threat model, risky operation description *ROD*, and using some NLP tasks such as textual entailment and dependency parsing. As shown in Fig. 4.4, Atomic works with a *Hazard Indicator Detector*, *LTE testcase generator* (LTCG), and *PoC Identifier* (PI). The HID extracts conditional statements, extends RODs, and runs a text entailment model to find HIs from all located conditional sentences, those implying any of the events and RODs. The LTCG analyzes the semantics of each HI to discover the state and events described and based on such information, constructs the testcases—a message to be issued by the adversary at the state through the templates retrieved from a database. The PI then executes the testcase on a configured simulation environment and automatically analyzes its logs to determine whether the attack succeeds.

4.3.3.2 Limitations and Discussion

All approaches prior to Atomic required significant human effort in documentation inspection, which is expensive and error-prone. Therefore, the approach utilized by Atomic is interesting as it takes the first step toward automating such tasks. In spite of this, the design of Atomic is still limited to HIs that include a single sentence or within a well-formed multi-sentence structure, and the risky operation of the HI is described as a verb phrase. Most of the issues found by Atomic are related to Denial-of-Service (DoS), because operations and HIs related to DoS are more explicit in the standards.

4.3.4 Semantic Information Analysis of Developer's Guide

Chen et al. [47] propose an approach for using the documentation of emerging syndication services to detect whether these contain logical vulnerabilities. They look at whether a syndication service will cause some security requirements to become unenforceable due to losing visibility of key parameters for the parties involved in the syndication, or bring in implementation errors when required security checks fail to be communicated to the developer. To address such a goal, Chen et al. develop a suite of NLP techniques that enables automatic inspection of the syndication developer's guide, based on the payment models and security requirements from the payment service. In this regard, they develop Dilution, a technique that automatically analyzes the developer's guide of payment syndication to infer potential security flaws in the merchant system. Dilution uses NLP to automatically uncover semantic information from the wrapper's integration instructions documented by the developer's guide and compare it against the FSM of the payment service encapsulated that is manually extracted. Figure 4.5 shows a high-level overview of Dilution.

4.3.4.1 Preprocessing

Though Dilution is claimed to be an automated tool, it requires a lot of manual preprocessing. For instance, it involves a one-time preprocessing step in which the FSMs of major payment processors are constructed and security requirements (SRs) for different states are identified. Such information is extracted from documents. Usually, payment process is clearly described through a diagram, which can be converted into an FSM.

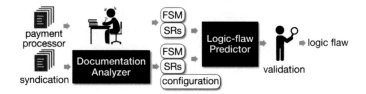

Fig. 4.5 Overview of Dilution [47]

4.3.4.2 Document Analyzer

Here, the FSM and SRs are used to automatically analyze the developerâŁ™s guides of different syndication services. The most important part here is to discover the syndicated payment process by extending the payment FSM, which further allows one to find out whether the parameters of required security checks are still visible to the new states supposed to perform the checks, and how these security requirements are explained to the developer. To achieve this, Dilution utilizes a suite of NLP techniques to first find out all sentences related to transmission activities, then performs syntactic analysis on each sentence and further converts detected syntactic elements to a semantic triplet describing the parties involved in a transition as well as the message sent.

4.3.4.3 Logic-Flaw Predictor

From the FSM and the SR information discovered from the syndication documentation, the logic-flaw predictor infers the possible presence of logic flaws, that is, the SRs expected by the payment processor that cannot be fulfilled in the syndication FSM, and those that have not been explained to the developer. To achieve this, Chen et al. define two types of security goals: (1) secure design; (2) secure implementation. The idea of a secure design is that for any security requirement to be enforced, there should be an enforceable security requirement in any state. Intuitively, this means that every payment security requirement should be fulfilled after syndication. Secure implementation on the other hand means that every security requirement is correctly implemented by either the merchant or the syndicator. For finding these, Dilution goes through the developer's guide and the extended FSM for the indicators that could lead to implementation flaws. The most important one is the absence of explanations about security requirements, a clear signal that the related security checks might not be implemented by the uninformed developer.

4.3.4.4 Limitations and Discussion

This research demonstrates that logic flaws or issues in the implementation can be predicted from documentation or standards. However, though the approach is claimed to be automatic, there is a lot of manual expertise and effort involved. Particularly, the extraction of SRs and FSMs is carried out manually. Though SRs and FSMs are small in number, we feel with the recent improvements in the state-of-the-art NLP tools, this can also be automated. The ultimate goal would be to have a completely automated analyzer that can analyze and specification for several downstream tasks such as policy enforcement weakness and inaccurate guidelines.

4.3.5 Security-Specific Change Request Detection

Chen at al. [45] have proposed an approach to understand security hazards in cellular network protocols (e.g., 5G) by an NLP analysis of Security Related Change Requests (SR-CRs) from the available Change Requests (CRs). They developed a pipeline called CREEK (CR Seeker). This approach is innovative in that it uses state-of-the-art language models such as BERT for cellular network analysis. Figure 4.6 shows an overview of the approach.

4.3.5.1 Embedding Generation
In this step, every sentence in each CR is converted into an embedding, a feature vector of the same size that captures the key information of the input sentence with a varying length. For this, BERT [63] is used. Then the pre-trained BERT is fine-tuned on all 3GPP specifications using two tasks, namely masked language modeling and binary classification for security-related specifications.

4.3.5.2 PU Learning
For the embedding generated by Step 1, high-quality labeling for training a classifier is needed. As there is only a small set of positive instances, they use positive-unlabeled (PU) learning to build the classifier. They develop an adversarial learning framework [105] with a classifier C and a discriminator D: D tries to recover the bias between the training distribution and the testing distribution, while C seeks an optimal separation between positive instances and negative ones with sample weights calculated from the bias recovered by D.

4.3.5.3 Self-Training
The last step is training. After the classifier is trained on the 10% unlabeled data, an uncertainty-aware self-training algorithm is run on the remaining 90% of the data to refine the classifier. Finally, the trained classifier is used to detect CR-SRs.

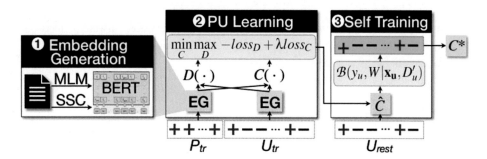

Fig. 4.6 Overview of CREEK [45]

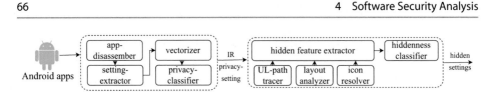

Fig. 4.7 Overview of Hound (redrawn from Fig. 5 in [48])

4.3.5.4 Limitations and Discussion

This is the first approach that uses large language models such as BERTs for the security analysis of cellular network protocols. However, among all the SR-CRs discovered in the author's research, 453 have not yet been accepted. An in-depth analysis of the ecosystem of these SR-CRs is an important issue that the authors do not consider. Therefore, they have some false positives. Future research needs to revisit these problems, and this can be an important research work.

4.3.6 Capturing Privacy-Related Settings in Android

Mobile apps include privacy settings that allow their users to configure how their data should be shared. These settings, however, are often hard to locate and hard to understand by the users. To address such a problem, Chen et al. [48] carry out the first large-scale measurement study on such hidden privacy settings. They develop an automatic analysis tool, called Hound, to recover unique features from the app and detect hidden settings. In the tool, they utilize NLP to capture privacy-related settings, by training a classifier on top of a set of feature vectors, each constructed from a settings' description. From these settings, Hound further discovers those considered to be hidden, based on the features identified. To be frank, this is a very rudimentary use of NLP. Figure 4.7 shows an overview of the approach.

4.4   Supporting Techniques

In what follows, we discuss two important tasks that are critical for software security analysis, for which ML approaches have been proposed. They are function boundary and type identification in binaries, and reverse engineering.

4.4.1 Neural Network-Based Function and Type Identification

Binary analysis facilitates many important applications like malware detection and automatic patching of vulnerable software. Some of the important problems in binary analysis

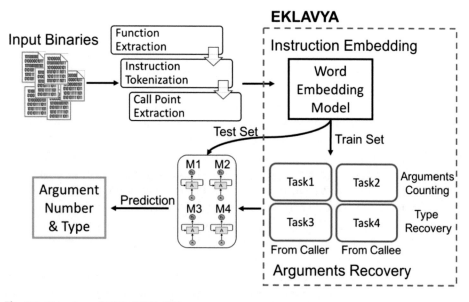

Fig. 4.8 Overview of EKLAVYA [50]

are function type signature identification, and function identification. Here, we discuss two approaches that use neural networks for binary analysis: one that recognizes functions in binaries [204], the other that learns function type signature dubbed as EKLAVYA [50]. The approach by Shin et al. [204] is based on training a recurrent neural network (RNN) to take bytes of the binary as input, and predict, for each location, whether a function boundary is present at that location. For this, no preprocessing is needed, such as disassembly or normalization of immediate operands in order to obtain good results. The EKLAVYA approach also uses a RNN architecture to learn function types from disassembled binary code of functions. One of the most important requirements in the design of EKLAVYA is explicability, that is, to find evidence that the learning network could learn something explainable or comparable. To gather evidence on the correctness of a learning network's outputs, Chua et al. [50] employ techniques to measure its explicability using analogical reasoning, dimensionality reduction, and saliency maps. By using these techniques, one is able to select the network architecture that exhibits consistent evidence of learning meaningful artifacts. See Fig. 4.8 for a high-level overview of EKLAVYA.

4.4.2 Reverse Engineering

In security-oriented software investigation and reverse engineering tasks that involve binary analysis, one of the most critical steps is to disassemble the target binary. Disassembly aims to recover the high-level structure from a binary code, such as assembly code, function, data types, and control flow information. However, due to compilation, optimization, and stripping, this type of high-level information is not accessible in most binaries, and disassembling tools often resort to approximations, which are frequently inaccurate. The problem worsens in the case of obfuscated binaries. Therefore, Pei et al. [179] propose to use TL to aid binary disassembling tasks and implement a disassembler called XDA based on their intuition. In particular, XDA focuses on recovering function boundaries and assembly instruction boundaries. To do so, they adopt a 2-step TL approach (based on BERT [63]). In the first step, XDA learns to "understand" machine code semantics from raw bytes. Using masked language modeling, XDA learns to recognize randomly masked bytes from byte sequences based on the correlation with the other bytes in the sequences. In the second step, XDA is fine-tuned to recover functions and instructions in byte sequences by recognizing their boundaries. The entire process is based on unsupervised learning and, thus, completely automated. For both steps, Pei et al. suggest employing self-attention layers as they capture the long-range dependencies in byte sequences better than sequential connections in RNN. XDA shows solid results achieving high accuracy (>99 F1) in recovering functions and assembly instructions from binaries compiled with several compilers. The results significantly outperform existing tools and previously proposed methods based on bidirectional RNN [204].

Prior to XDA, Wang et al. [240] proposed FID, an approach to automatically recognize functions in binary code based on classification. Their approach is unique in that it combines disassembling, symbolic execution, and ML classification. FID first employs an open-source reverse engineering tool UROBOROS [241] to disassemble the target binary and produce the program control flow structures, such as the basic blocks. Then, FID applies symbolic execution to each basic block to generate accurate semantics of the blocks. As the symbolic engine executes each basic block, FID records information such as assignment formulas for general-purpose registers (e.g., esp and epb) and memory accesses. The authors' intuition is that stack memory operations involving registers are often related to function parameter read operations, which can help detect the functions' entry point. To support effective learning, FID needs to translate the semantics into numeric feature vectors. Thus, FID first selects the most informative semantics (e.g., stack registers and memory read operations) and then converts them into numerical vectors by leveraging lexical and syntactic features, such as the number of tokens in a formula divided by the length of characters (lexical) or the maximum level of nested parenthesis (syntactic). Subsequently, FID employs three classification algorithms, namely LinearSVC, AdaBoost, and Gradient Boosting, to implement a majority voting mechanism to classify the disassembly instructions and detect functions' entry points. Finally, FID splits the code into multiple regions according to the functions' entry point and reconstructs the code control flow graph (CFG) [21] to recover the functions' boundaries.

4.5   Research Directions

The application of ML to software security has several open directions, many of which have been discussed throughout the chapter. Here, we mention two additional directions. The first is related to the ML-based enhancement of static analysis; the chapter has described some recent approaches. However, more work is needed to develop ML-based approaches for detecting specific types of vulnerability. The second is related to fuzzing and is based on the observation that the ML-based approaches that have been proposed focus each on using ML for one specific phase in the fuzzing process. It would be interesting to evaluate the efficiency and accuracy of fuzzing when using ML for all the phases or a subset of the phases.

Hardware Security Analysis

Hardware is the backbone of all computing resources, and its security is of great importance. Compared with software vulnerabilities, hardware vulnerabilities are much harder to deal with because hardware cannot be patched easily as software. Thus, hardware developers heavily rely on hardware verification to find vulnerabilities in the prototype before the official release. In fact, in the hardware design cycle, hardware verification can take 60–75% of the time, computing, and human resources.

Traditional approaches for hardware verification can be divided into two categories: formal verification and random testing. Formal verification aims to mathematically prove that the hardware implementation meets the desired specification. However, formal verification requires substantial manual and expertise efforts and does not scale as the hardware evolves quickly. Thus, random testing—a more practical method that uses various randomly generated inputs to exercise the hardware—has recently become more popular.

However, random testing has the drawback that it does not guarantee testing completeness. Due to the hardware complexity, certain parts of the hardware can only be tested with dedicated test inputs, which can hardly be generated randomly. For instance, testing a certain circuit block might require test inputs that satisfy a very specific value. As a result, hardware developers still need to spend a lot of manual effort writing heuristics-based rules to limit the search space of the new inputs and therefore increase the chance to find hard-to-find test inputs.

Several ML approaches have been recently proposed to avoid such manual efforts. Intuitively, a ML model can be trained to learn the hardware executions that are triggered by different test inputs, and then be used to help generate test inputs that trigger uncovered execution paths. Some approaches use ML to guide the input generator generating unique inputs that cover different parts of the hardware. Others use ML to perform binary classifi-

© The Author(s), under exclusive license to Springer Nature Switzerland AG 2023   71
E. Bertino et al., *Machine Learning Techniques for Cybersecurity*, Synthesis Lectures on Information Security, Privacy, and Trust,
https://doi.org/10.1007/978-3-031-28259-1_5

cation on randomly generated test inputs into error-inducing and legitimate inputs, so that only error-inducing inputs need to be tested and analyzed.

Another important verification task for hardware security is related to the detection of hardware Trojan. A hardware Trojan is a malicious modification of the circuitry of an integrated circuit. The problem of hardware Trojan has been known for a long time [42], but it still continues to be of major interest. Therefore, recent work has leveraged ML techniques to design more effective approaches in hardware Trojan detection.

In what follows, we first present approaches that use ML for enhancing test input generation. We then focus on the use of ML for detecting hardware Trojans.

5.1   ML-Based Hardware Test Input Generation

Random testing on hardware is mainly performed by simulation. Hardware developers first describe the functionality of the hardware in specialized languages such as Verilog RTL. Next, they can use automated tools to randomly generate test inputs to the hardware, during which certain constraints or rules can be applied to guide the input generation so that the hardware can be tested thoroughly with distinct inputs. Then the test inputs are provided to the hardware under simulation to test if the expected results are obtained. Any simulation failures (e.g., incorrect results) indicate that the hardware design is problematic and should be fixed.

Additionally, because passing a certain number of test inputs does not necessarily mean that the hardware design is free from bugs, hardware developers often use metrics to measure the amount of effective testing that has been performed. And the testing only becomes sufficient when the target coverage is fully achieved. For instance, the hardware is only considered to be fully tested if all unique hardware states are tested by at least one test input.

Recently, ML is being used to guide the input generation for hardware [84, 114, 115] to improve the coverage (testing completeness) or efficiency (finding bugs quickly). Existing approaches use different ML models but share a similar workflow, consisting of the following steps:

1. Collect a set of test inputs as training samples. Example inputs can be generated randomly with existing manually written constraints and rules. They can be tested individually under simulation to collect the labels. In practice, the coverage triggered by a test input will be used as the label.
2. Train a ML model on the training samples. Generally, the input to the model is the test input, which is usually expressed as a byte or text sequences. The output of the model is the coverage, often expressed in binary vectors. The relationship between test inputs and the coverage they trigger is learned by the model.
3. Use the model to assist with hardware verification tasks. For this task, some approaches use the trained model to predict the possible coverage for a given test input. Intuitively,

if the prediction indicates that the input will not contribute to increasing the overall coverage, it is likely that this input is redundant and should be de-prioritized or even skipped. Other approaches use the model to generate new test inputs. As the model learns the function mapping from a test input to the coverage result, certain computations can be performed to know which parts of the input should be changed in order to cover a new state or condition.

Case-1: Use ML to generate effective test input for hardware verification

Design2vec is a recently proposed ML architecture for learning continuous representations to capture the semantics of a hardware design [234]. Design2vec (see Fig. 5.1) enables several key hardware verification tasks such as coverage prediction and test input generation assistance. The insight behind the Design2vec approach is that the hardware design can be abstracted at a higher level than gates or bits, and can thus be described using digital circuit languages, such as the Register Transfer Level (RTL), which is similar to software source code. The benefit of using RTL is that it allows one to use software representation techniques for hardware-related tasks. For instance, a circuit described in RTL can then be represented in a graph that incorporates control and data flow, which then can be learned effectively using graph neural networks.

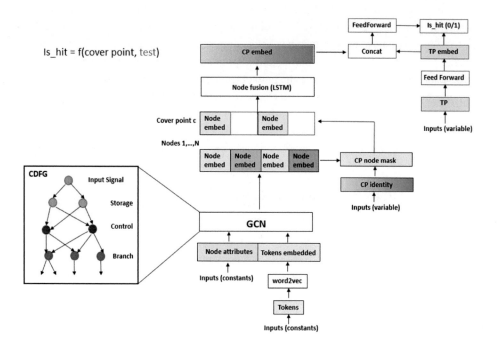

Fig. 5.1 Overview of the Design2vec model (redrawn from Fig. 2 in [234])

Design2vec constructs a graph for the hardware described by RTL and then uses a graph neural network (GNN) to learn the hardware semantics. To build the graph, Design2vec first represents each statement in the RTL code, the feature of which contains information, such as the node identifier, node type, and the statement token sequence. Nodes are connected by heterogeneous edges such as control and data flow edges. In particular, hardware execution, unlike software execution, has a unique non-terminating property. For instance, if the execution in the current cycle reaches the end of a circuit block, in the next cycle the execution will continue from the beginning of the block, instead of exiting the block as the software execution. To reflect this property of the hardware, Design2vec adds artificial control edges that connect the last and the first statements in each code block, indicating that the code block will execute in a dead loop.

The graph representation of the hardware, as well as the test input, are the input to the model to perform learning and prediction. For the graph input, Design2vec first performs an embedding of the node feature using LSTM. Then it uses a new model architecture, called RTL IPA-GNN, to learn the graph. For the test input, Design2vec uses a feedforward multi-layer perceptron (MLP) layer for the embedding. The embeddings of the graph and the test input are finally concatenated and then provided to a feedforward layer (with sigmoid activation) to predict each code block (e.g., whether this block will be covered or not given the test input).

Design2vec has been evaluated on real-world hardware including IBEX v1, IBEX v2, and Google TPU for two tasks. In the task of predicting hardware input coverage, Design2vec can achieve an accuracy as high as 90%, indicating that it can be used in practice. In the task of test input guidance, Design2vec has been compared against Vizier SRP [87], a random testing tool that uses black-box optimization to maximize total coverage. Specifically, the experiments show that Design2vec can both (1) find more hardware failures, in that it is able to generate inputs that can trigger failures that Vizier cannot find; (2) find failures quickly, in that it is able to trigger a hardware failure with fewer test inputs generated and tested than Vizier.

Case-2: Use ML to identify error-inducing hardware input

A recent work [211] by ARM researchers has explored the use of ML techniques to identify test inputs that might cause hardware failures. Specifically, they propose a testing workflow where both traditional random-based testing and ML-guided testing are executed at the same time (see Fig. 5.2). The random-based testing keeps generating various test inputs to exercise the hardware, during which a large amount of data is generated and then used to train ML models. The ML-guided testing uses an ensemble of both supervised and unsupervised models to flag test inputs that might cause hardware failures. A test input is flagged as suspicious and is tested later if at least one model predicts it as suspicious.

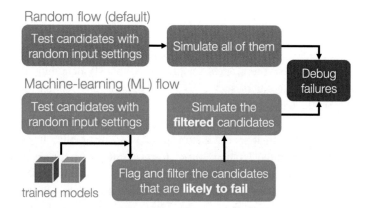

Fig. 5.2 Overview of the approach using ML for error-inducing test input [211]

In terms of the model architecture, several non-neural-net classifiers, such as logistic regression and random forest, have been evaluated as part of this work, whereas isolation forest has been used as an unsupervised learning algorithm. The models were trained on a dataset collected over a month, which contains many test inputs and their execution results.

The evaluation shows that the ensemble model is promising. First, the results show that the model helped find 80% unique hardware failures by running only 40% of test inputs. Furthermore, the approach shows a relatively high recall score (e.g., 85%) and high efficiency in the metrics adopted by the authors of the approach. For example, under the metric that divides the number of unique bugs found by the number of unique failures predicted to happen, the model can achieve scores as high as 80%.

5.2   ML-Based Detection of Hardware Trojans

Hardware Trojans are malicious modifications introduced into hardware design by an adversary. Identifying hardware Trojans is essential for securing hardware given the recent evolving trends of the semiconductor supply chain that involves many entities [108]. Techniques for the detection of hardware Trojans are used to check whether unwanted circuits are implanted in designed or fabricated integrated circuits. Toward such a goal, ML-based techniques have been proposed with respect to

- Reverse engineering, which aims to repackage the integrated circuits to reconstruct the original design. Proposed approaches include (i) using SVM and k-means to create models able to distinguish between the expected and suspicious structures on integrated circuits [22, 23]; (ii) creating an input-output model of the integrated circuits via behavioral pattern mining and then discovering Trojans via behavioral pattern matching [120].

- Circuit feature analysis, which aims to extract circuit features, either functional or structural ones, and identify whether a net or gate is suspicious. The goal is to identify Trojans that are activated under rare conditions or implanted into the design of integrated circuits as redundancy modules. The proposed approaches typically determine the features to extract and then train classifiers, such as SVM and ANN, to differentiate circuits infected by Trojans from the normal ones.
- Side-channel analysis, which measures circuit parameters, such as power, path delay, temperature, and electromagnetic radiation profiles, to distinguish integrated circuits infected by Trojan from the "golden" ones. However, due to the high complexity, side-channel traces tend to contain a lot of noise (resulting from process variances) and make manually crafted analysis rules ineffective. Side-channel analysis has attracted a lot of interest in the past few years and recently a few approaches have been proposed that leverage ML to address issues related to noise.

In what follows, we summarize two recent approaches for side-channel analysis. We refer the reader to the survey by Huang et al. [108] for a detailed discussion of ML-based approaches for protection against hardware Trojans.

Yang et al. [251] explore different ML techniques to detect potential hardware Trojans. The key insight is that the side-channel signature can be analyzed using statistical methods. They built a dataset on a custom hardware platform (HaHa board) and evaluated ML techniques such as naive Bayes, decision tree, K-nearest neighbor, and SVM. Models are then trained to detect Trojans of different types, sizes, and chip locations. The results show that SVM is better than other ML models in terms of performance and scalability.

Ashok et al. [19] (see Fig. 5.3) focus on finding better inputs that have more information for the ML models. They propose to use Quantum Diamond Microscope (QDM) to generate high spatial resolution magnetic fields of the hardware chip. QDM uses internal quantum transitions of optically active nitrogen-vacancy centers in a chip so that it can provide a more sensitive and high-resolution image of the chip. Therefore, QDM allows one to collect very precise measurements. They apply some CNN models to learn the QDM image. For every

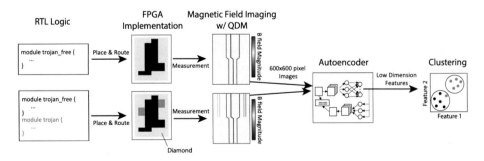

Fig. 5.3 Overview of the approach using ML for better hardware representation [19]

graph, the CNN model outputs a vector representation. Then a clustering algorithm, such as DBSCAN, is applied to all graphs to cluster them into Trojan-free and Trojan-infected groups according to their vector representations. Such an approach has the advantage that it does not require comparison with the golden circuits to detect Trojans. The evaluation shows that the Trojan detection accuracy is over 87%. Additionally, they compare the CNN-based approach against PCA—a traditional technique for processing vector representations—and found that using the CNN increases the accuracy by 10%.

5.3   Research Directions

Hardware security analysis is an area where interesting ML-based approaches have been proposed. However, it is clear that much more work is required to understand how best to leverage ML techniques and how to extend these techniques for hardware security. An interesting direction is to develop a detailed taxonomy of vulnerabilities and identify which ML techniques are more suitable for each such vulnerability. Also, ML techniques should be analyzed with respect to the hardware life cycle to identify for example approaches that can be used to detect anomalies at runtime.

Detection

6

Intrusion detection represents a fundamental security function. It consists of monitoring network traffic and host processes for suspicious activity, and issuing alerts when such activity is discovered.

There are two main approaches to intrusion detection: signature-based detection and anomaly detection (AD). In the former approach, a signature is a pattern of a specific intrusion. Once the system detects on-the-fly any pattern that exactly matches a signature, it raises an alarm. For instance, let us assume that there is an attack packet that always contains a special character "$" in the User-Agent field in the HTTP header. The detection system raises an alarm if it sees a packet exactly matching the signature, which is "'$' in the User-Agent field in the HTTP header". Although this approach has a high true positive rate on known attacks, the main drawback is that it cannot detect unknown attacks. For example, the detector cannot detect an attack variant packet that contains "@" instead of "$" in the User-Agent field in the HTTP header. To detect a new attack, a human should understand the new pattern, generate the corresponding signature, and add it to the set of signatures of the detector. On the other hand, AD is effective in detecting unknown attacks. This approach learns the normal behavior pattern of a system and raises an alarm if a given pattern deviates from the normal pattern. The main advantage of AD is the ability to automatically detect unknown attacks. However, it is important to point out that combining both types of detection provides a more comprehensive detection-based defense.

Another important category of detection is related to malware detection and analysis, which is typically executed to determine whether a piece of software is malicious and if so to determine the type of the malware. The latter is critical to understand, for example, if a previously unseen malware is derived from some known malware. If this is the case, one can often eradicate the new malware by following the same steps taken for the known

© The Author(s), under exclusive license to Springer Nature Switzerland AG 2023
E. Bertino et al., *Machine Learning Techniques for Cybersecurity*, Synthesis Lectures on Information Security, Privacy, and Trust,
https://doi.org/10.1007/978-3-031-28259-1_6

malware. Approaches for malware detection and analysis are mainly based on analysis of the malware code. However for certain types of malware, such as ransomware, one may also use behavioral analysis, based on detecting specific anomalies in the execution of processes.

ML techniques have been widely used in both categories of detection. In what follows, after an introduction of the main malware categories, we cover ML-based anomaly detection and then ML-based malware detection and analysis.

6.1   Types of Malware

There are several malware families that typically differ with respect to their goal and approach:

- Adware: It is a form of malware that hides in user interfaces and serves advertisements to the users. Some adware also monitors user behavior online so it can target users with specific ads.
- Fileless Malware: It is different from other malware in that it does not rely on files containing malicious code to infect a host. Instead, it exploits applications that are commonly used for legitimate and justified activities to execute malicious code in resident memory. This malware is hard to detect as it attaches to legitimate scripts, and these scripts continue to run with the hidden malware.
- Viruses: These are malicious executable codes that usually attach themselves to an executable file and get triggered when the infected file is opened. The virus exploits vulnerabilities of the operating systems; sharing of files spread the virus across systems.
- Worms: They act like viruses that leverage security loopholes and infect the system by replicating themselves. The worms do not require any human activity to propagate as they are self-replicating.
- Trojans: This type of malware seems like legitimate software and tricks the users to download and install the malicious software into their systems. Once a Trojan enters the victim system, it can perform all the tasks that legitimate users can perform, like reading, modifying, deleting, and transferring files, connecting over the Internet, and redirecting the traffic.
- Botnets: They are networks of interconnected infected devices that are managed and run by a command-and-control server. The infected devices, often referred to as bots, are tasked with executing various operations to conduct malicious activities, like performing a distributed DoS (DDoS) attack.
- Ransomware: The ransomware makes files unavailable, by typically encrypting them with an encryption key not available to the victim, and demands a ransom in order to make the files available again to legitimate users (see also the discussion in Sect. 6.2.4).
- Downloader: It is malicious software used to support other malware. The purpose of the downloader is to download additional malicious files needed for the malware on the infected system.

6.2   ML-Based Anomaly Detection

Detecting behavioral anomalies in computer systems is an essential step in many defenses. The main objective of AD is to learn the *normal behavior* of the system over some period of time, and then flag any activity that deviates from that learned behavior. In many cases, learning the normal behavior requires monitoring and recording all the activities being executed in the system in order to extract a pattern or a profile for each user/application. Due to their broad applicability, anomaly detectors have been successfully applied in many domains such as network management [2, 76] and operating systems [217].

Anomaly detectors typically consist of two modules, namely a *data collector* and a *data analyzer*. The former is a module that collects logs from the system being monitored, while the latter is a module that analyzes the collected logs. Depending on the placement of data collectors, there are two types of anomaly detectors: network-based ones and host-based ones. Those two types of detectors generate different types of logs; thus, features used for anomaly detection are different, depending on the placement of the detectors.

Over the collected logs, a data analyzer performs AD. In what follows, we discuss *what algorithm* is used to address *what problem*. The algorithms are categorized as supervised, semi-supervised, and unsupervised techniques. Within each category, there are many different algorithms, such as LSTM or autoencoder (see Chap. 2). Furthermore, the features one can use for AD are diverse, depending on the types of logs. For instance, from network packets, a detector can use as features inter-arrival time between packets or the number of packets in one direction. From host logs, a detector can use CPU/memory utilization. Also, the features may differ depending on the specific threat against which one deploys the anomaly detector. For example, an anomaly detector for ransomware detection (see Sect. 6.2.4) uses features that are different with respect to the features used by an anomaly detector for cryptojacking detection [226].

There are several challenges when deploying an AD system. A major challenge is how to obtain the normal behavior patterns. It is difficult to define what "normal behavior" is; thus, it is challenging to collect such data. Another challenge is that a detector may suffer from high number of false positives as it may raise an alarm for all the unexpected patterns. Unfortunately, unexpected patterns do not always mean that there is an intrusion. Rather, there can be lots of unexpected normal patterns, for example, due to congestion in networks or bottlenecks in systems. Therefore, reducing the number of false positives is critical for the successful deployment of AD systems.

In what follows, we discuss AD approaches proposed in diverse domains including networks, IoT systems, and cyber-physical systems (CPS). For each domain, we discuss challenging issues and how ML algorithms help to address them. We also describe an AD system for ransomware as this gives an example of specific features required for AD that are different from features typically used in AD for networks and hosts, as selecting the proper features is a critical step for AD.

6.2.1   Networks

A network-based AD system is a detector in which a data collector is placed on an edge of a network the detector aims to protect. For instance, a data collector is typically located on a gateway and collects packets passing through the gateway. Based on the packets, a data analyzer performs AD. In what follows, we describe reconstruction-based approaches proposed for learning the normal behavior of networks. Then, as networking patterns are usually time-dependent, we describe approaches using AD based on multivariate time series. Finally, we describe one interesting approach that uses DL for feature engineering.

6.2.1.1 Reconstruction-Based Approaches Through Zero-Positive Learning

The reconstruction-based approach relies on a model that learns a normal distribution. A given pattern should be reconstructed by the model if it follows a normal distribution. Otherwise, the output of the model would significantly deviate from the input, which turns out to be an anomaly.

One of the example systems is Kitsune [158], an AD system that uses an ensemble of autoencoders. To avoid the expensive labeling process typical of supervised learning, Kitsune uses as a model an autoencoder because it is effective for unsupervised learning. Furthermore, Kitsune has been designed for deployment at routers and therefore being lightweight is a critical design requirement. Once Kitsune acquires a packet, it parses the packet to get the meta information and then sends it to the feature extractor, which extracts features and generates an instance. The feature extractor obtains temporal statistical features using incremental statistics for small memory. Next, based on the extracted features, the feature mapper splits the instance into several sub-instances, each of which is fed to the ensemble of autoencoders. The main reason for this splitting is to make sure that AD is executed effectively with low complexity. Finally, the ensemble outputs whether a given packet is anomalous or not.

DAGMM [257] is another example of an anomaly detector that uses an autoencoding Gaussian mixture model. It performs density estimation to determine anomalies, after reducing the dimensionality of the given samples. However, important features for density estimation may be lost during dimensionality reduction; thus, an important requirement is to avoid losing important features for better performance. To this end, DAGM leverages an autoencoder that extracts useful features as its non-linear dimensionality reduction helps to get important features, represented as a low-dimensional sample (i.e., a latent representation), and concatenates them with reconstruction error features, such as Euclidean distance. Then the resulting sample is fed to the detector that performs density estimation under the Gaussian mixture model to determine anomalies.

ZeroWall [224] is also an example anomaly detector that aims to detect zero-day attacks on web requests based on unsupervised encoder-decoder recurrent neural networks. The main insight, on which ZeroWall is based, is that benign web requests follow the HTTP protocol while malicious web requests show inconsistent syntax and semantic patterns. ZeroWall augments existing signature-based firewalls so that it is immediately deployable. By being co-located with existing firewalls, it benefits from collected web requests allowed by the firewalls as benign training sets based on the observation that zero-day attacks rarely happen. Then, it applies machine translation quality assessment techniques to address the detection problem. In detail, ZeroWall trains an encoder-decoder model that learns from benign web requests as language. By tokenizing the requests and generating a sequence of resulting tokens, the model generates an LSTM encoder that encodes a given sequence into a latent representation, and an LSTM decoder that outputs a reconstructed sequence from the latent representation. Anomalies are detected if the model cannot reconstruct a given sequence.

6.2.1.2 Multivariate Time-Series-Based Approaches

A multivariate time series is a series with multiple time-dependent variables. It is useful in AD as recent attacks typically follow multiple steps to achieve their goals and time series are helpful to understand the attack contexts for detecting anomalies.

MTAD-GAT [255] uses not only multiple features in its decisions but also the temporal dependencies and the correlation between different features. It is different from other approaches that analyze each variable independently. By correlating different features, MTAD-GAT can detect unexpected but normal patterns, reducing the number of false positives. The rationale behind it is that a sudden change in a certain metric does not always mean that the change is actually indicative of an intrusion. For example, the CPU utilization of a server can abruptly increase because of the normal behavior of a particular process. If using only the CPU utilization metric, the system would always raise an alarm, resulting in lots of false positives. However, by considering the information about the process simultaneously when the CPU utilization has skyrocketed, the system may understand normal cases with sudden changes. To this end, MTAD-GAT introduces the graph attention layer to detect multivariate correlations. With the graph attention layer, MTAD-GAT performs AD through a combination of single-timestamp predictions and reconstruction of the entire time series.

6.2.1.3 Learning Without Feature Selection

nPrint [110] is a unified feature vector that captures diverse types of packets, namely TCP, UDP, and ICMP. As such packets have different header structures, nPrint represents packets based on all the header fields in TCP, UDP, and ICMP. For example, nPrint has a sequence number field (TCP). When a TCP packet should be described in nPrint, a sequence number in the packet is assigned to the field. However, a UDP packet does not have a sequence number; thus, nPrint assigns -1 to the field. In this way, all the IP packets can be vectorized regardless of the transport layer protocol. Based on the vectors, a ML algorithm [97] is used to select an appropriate model with the best hyperparameters. The experiment result shows that the performance of the model achieves high accuracy on the netML challenge [24], which consists of flow-level traffic analysis problems, namely malware detection for IoT devices, intrusion detection, and traffic identification. The result is promising as it removes the need of selecting the proper features to be used for ML-based AD and thus reduces human costs.

6.2.2   IoT Systems

IoT systems are increasingly being deployed in many application-critical sectors, and therefore several ML-based intrusion detection techniques have been proposed for their protection. The design of those techniques requires, however, addressing several challenges, including the limited computing resources of IoT devices and the lack of sufficient training data. Also, IoT systems often use wireless communication networks with specialized protocols and therefore IDSes must be customized to detect attacks on these networks. In what follows, we first introduce two approaches to deal with the limited computing resources of IoT devices, and then an approach, based on federated learning, to deal with the lack of sufficient training data. We then conclude with approaches specialized for detecting attacks on wireless networks used by IoT systems.

6.2.2.1 Architectural Solutions for Constrained IoT Devices

Deploying a host-based anomaly detector on IoT devices is challenging due to their limited capabilities. To address this challenge, the E-spion intrusion detection system [163] has been designed with a device-edge split architecture where data collectors are located on IoT devices, while the data analyzer, based on ML, is placed at an edge server. As the data analyzer performs the heavy computation, the IoT devices only have to perform minimal work, which consists of collecting logs and sending them to the data analyzer. An interesting aspect of E-spion is represented by three different levels of granularity for data collection at the IoT devices. At the lowest level, the log only collects the names of processes running on the devices. At the medium level, the log collects metrics, such as CPU and IO usage. At the highest level, the log collects the number of system calls, for each type of call. The highest level provides the highest accuracy while however increasing the processing costs at the devices, whereas the lowest level has the lowest accuracy but incurs the lowest processing costs at the devices. E-spion can then be configured by choosing the data collection granularity which represents the best trade-off for the different devices in the IoT system of interest.

6.2.2.2 Selection of Appropriate Detectors

Pasikhani et al. [177] propose a network-based, hybrid-based, and distributed IDS based on RL, referred to as RL-IDS. The proposed IDS employs several lightweight signature-based or anomaly-based detectors, each of which is built from a subset of the training dataset. The subsets have different patterns of attacks; thus, the detectors have different strengths and weaknesses in analyzing different attacks. RL-IDS then uses RL to select an appropriate detector to analyze the current input packets. The agent gets a reward when it successfully selects the best detector for particular scenarios.

6.2.2.3 Federated Learning Techniques to Deal with Insufficient Data

As networking patterns of IoT devices are usually simple and scarce, it is challenging to learn the normal behavior of them. To address such a problem, Nguyen et al. [171] propose DÏoT that leverages federated learning. Motivated by similar patterns of the same type of devices such as a camera, DÏoT collects data per device type and performs AD based on the type of device. In other words, the intrusion detection system (IDS) uses federated learning to generate a profile for specific types of devices, for which the data used for the profile are collected and aggregated across several networks. The resulting model is managed by a global IoT security service; thus, any DÏoT analyzer module can obtain the profiles from this service to carry out AD activities.

6.2.2.4 Detection of IoT-Specific Attacks

There are IoT-specific protocols supporting the connection to the Internet of devices with low capabilities. 6LoWPAN and RPL are two such protocols. 6LoWPAN is an adaptation protocol layer for low-rate wireless personal area network (IEEE 802.15.4 hereafter) devices that allow these devices to use IP version 6. As the size of the frame of IEEE 802.15.4 is only 127 bytes, IEEE 802.15.4 cannot use IPv6, which requires at least 1,280 bytes. 6LoWPAN addresses this limitation by using fragmentation, reassembly, and compression. On the other hand, RPL is a routing protocol between these devices. It relies on the notion of destination-oriented directed acyclic graph (DODAG) for routing. To generate the DODAG, devices exchange routing information. However, such an information exchange between devices with low capabilities opens a new attack surface. For instance, by manipulating this information, an attacker can perform routing attacks. Therefore, appropriate security mechanisms are required.

Shukla [203] suggests three IDSes, namely KM-IDS, DT-IDS, and Hybrid IDS utilizing both KM-IDS and DT-IDS that aim to detect wormhole attacks. KM-IDS relies on the k-means clustering algorithm, while DT-IDS leverages the decision tree algorithm. Those IDSes detect the attack based on the distance threshold between any two devices. If an IDS determines that the distance between two devices is higher than the threshold, the IDS raises an alarm. To determine the exact distance, ML algorithms are used. In the case of KM-IDS, devices are clustered based on their distance from the gateway. Then, the attack is detected if two communicating devices are in different clusters. In DT-IDS, the IDS calculates the average of distances between two devices and sets the result as the threshold. Hybrid-IDS combines KM-IDS with DT-IDS. It uses two threshold values set by KM-IDS and DT-IDS, respectively. Whenever KM-IDS detects an anomaly based on clusters, Hybrid-IDS compares the distance between the two devices with the threshold set by DT-IDS. Hybrid-IDS only raises an alarm if the distance is longer than the threshold. This approach removes the false positive cases in which two close devices in two different clusters are around the boundary of the clusters, thus improving accuracy.

Napiah et al. [165] introduce an IDS, called CHA-IDS, that leverages the 6LoWPAN compression header. The reason for using the compression header is that it contains routing information; thus, such information is helpful to detect routing attacks. It is different from prior IDSes that usually use the rank of devices as the main feature in identifying routing attacks. The simulation results show that such a new feature enhances the model's accuracy in detecting anomalies in the 6LoWPAN packets.

6.2.3 Cyber-Physical Systems

A cyber-physical system (CPS) is a computer system that controls a set of given physical components by interacting with sensors and actuators. Examples of CPSes are smart factories, industrial control systems, smart grid, and intelligent transportation systems. In a CPS architecture (see Fig. 6.1), the physical components are controlled by a CPS workflow. The *sensors* are responsible for interfacing between a physical world and a cyber world by measuring the status of the physical components, while the *actuators* actively manipulate the state of physical components. The sensors and actuators are directly connected to control systems, like programming logic controllers (PLCs), which obtain data from the sensors and send commands to the actuators. The control systems are again supervised by a comprehensive system, called the supervisory control and data acquisition (SCADA) system.

Luo et al. [149] discuss AD systems in CPS applications including industrial control systems (ICSes), smart grid, intelligent transportation systems (ITS), and aerial systems against the following three potential threats. The first is the tampering of sensors and actuators; in this threat, sensor data collected from physical components and data reported to control systems are different. Similarly, commands from the control systems and commands to physical components are different. The second is the reading and manipulation of packets

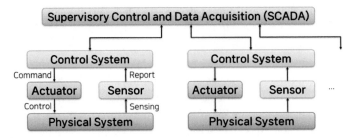

Fig. 6.1 A cyber-physical system (CPS) architecture

in the communication channels between physical components and sensors (or actuators), and between the SCADA system and the control systems. The third is the breaching of control systems. In other words, an attacker controls the control systems causing internal faults or false control signals.

A typical CPS AD system is designed considering three aspects. The first is the data used to train a model. Usually, sensor and actuator data, communication traffic, and control system logs are used. Those data are preprocessed and then fed to neural networks. Second, different ML models should be considered. Different models, such as RNN or autoencoders, can be used. Finally, metrics to update the models by loss functions that calculate anomaly scores referring to output data and ground truth data are one of the important factors. They are (1) prediction error, (2) reconstruction error, and (3) predicted labels.

Below, based on [149], we describe CPS AD systems in four CPS applications which adopt some novel approaches.

6.2.3.1 Industrial Control Systems

The area of AD for industrial control systems (ICSes) has been widely investigated. As CPS systems are usually resource-constrained, Feng et al. [78] have designed a system that uses the Bloom filter to capture suspicious packets, so that only these packets are transmitted to the ML model, thus greatly reducing the amount of packets that the model must process. Many systems leverage the temporal behaviors of data (i.e., time series) as a sensor's or an actuator's values are one-dimensional. However, there are relationships between sensors. Thus, approaches have been proposed [39, 116] that use CNNs to capture those correlations. A framework has also been proposed that leverages a GAN to capture the spatial-temporal correlation [147]. The framework detects false control signals from sensor and actuator values based on reconstructing discrimination errors from both the generator and discriminator.

6.2.3.2 Smart Grid

One of the most widely investigated attacks in the context of a smart grid is the false data injection attack. Such an attack aims to cause measurement errors to provide false information to state estimators, which leads to incorrect behavior of the smart grid. As the false data injection attack is typically stealthy, much research has focused on how to improve the accuracy of state estimators. One approach is to introduce a filter to remove the anomalous data items before transmitting the data to the state estimators. Basumallik et al. [25] have introduced a filter in the detection system that captures potentially anomalous data, which are then transferred to a detector. If potentially anomalous data is classified as anomalous data, it is not used in the estimation of the state of the system. Another approach is to improve AD by using joint detection. Wang et al. [242] have proposed a framework consisting of several autoencoders and conventional detectors. The framework raises an alarm only if both the autoencoders and the detectors raise the anomaly. This joint approach significantly reduces false positives.

6.2.3.3 Intelligent Transportation Systems

Most approaches proposed in this domain focus on the controller area network (CAN) bus, a communication bus between devices and electronic control units (ECUs). The threat model considered in this domain is an active network attacker on the CAN bus that performs traffic drop, reordering, replay, injection, and so on. For a vehicle, not only the sensor data in the CAN network is important to detect anomalies, but also the environment information is critical. The reason is that the same physical status can be classified differently according to the condition of the environment, such as the weather or the road. Kieu et al. [135] have proposed a method to capture context information by matrices, resulting in higher precision and recall. As the CAN network is delay-sensitive but usually resource-constrained, an approach has been proposed to boost computing relying on mobile edge devices [256]. Such an approach uses a multi-dimension LSTM framework computing part of which is delegated to mobile edge devices. An experimental evaluation shows that this approach significantly reduces the computation time (100 times faster). Robustness is also an important factor for intelligent transportation systems; thus, responses to environments should be correctly generated. To this end, Wyk et al. [233] have designed a two-level framework where the first level consists of a three-layer CNN model, followed by Kalman filters (KFs) in the second level. The KFs eliminate false sensor readings that are undetected by the first level; thus, this framework improves accuracy.

6.2.3.4 Aerial Systems

Several approaches have been proposed aiming to detect faults in aircraft or unmanned aerial vehicles (UAVs). As all approaches require a threshold to detect anomalies, setting up an appropriate threshold is important for the performance of the models. However, it is quite challenging to decide the accurate threshold as it needs to be set based on large number of observations. To address such a challenge, Hundman et al. [111] have proposed a framework to calculate the threshold in a dynamic and automatic way. In the first stage, the framework evaluates smoothed prediction errors. Then, an exponentially weighted average is applied to the errors. Next, the threshold is adjusted according to an expression that consists of the mean and standard deviations. Since spacecraft generate lots of telemetry data, the data size and noise can negatively affect the performance of the model. To address such an issue, Tariq et al. [225] have proposed a method to reduce the data amount by sampling. The method relies on ConvLSTM—a convolutional LSTM network, and mixtures of probabilistic PCA to detect anomalies.

6.2.3.5 Future Directions

Luo et al. [149] have identified several issues that need to be addressed in order to improve detection methods. First, as most approaches use empirical thresholds, which are not optimized and robust against changes in the environments, one important issue is to determine the threshold automatically and adaptively. Second, there is a need for benchmarks specific to CPS to assess the performance of AD techniques and systems. Third, the deployment of AD systems requires these systems to meet real-time requirements. Fourth, forensic analysis techniques are needed to identify the compromised devices, the steps of attacks, and the exploited vulnerabilities.

6.2.4 Ransomware

Ransomware attacks are today major security risks for public and private organizations around the world. In a ransomware attack, attackers remotely compromise computer systems, block access to the data by legitimate users—by typically encrypting the data with encryption keys only known to the attackers, and demand a ransom in return for restoring access to the data. More recently, attackers also steal the data and sell or expose the data if the victims do not pay the ransom, and have been targeting critical infrastructure [93]. As discussed in a recent report by the Institute for Security and Technology [113], this form of cybercrime leads to dangerous real-world consequences that far exceed the costs of ransom payments alone, including loss of human life.

Therefore, defenses against ransomware have been, and are still being, widely investigated. As discussed by Moussaileb et al. [160], defenses can be classified according to the typical workflow of ransomware attacks, namely infection, environment preparation, data access denial, and ransom request. Figure 6.2 lists examples of activities performed in each step—some of these activities may only be present in ransomware that not only encrypts the data but also exfiltrates the data to then sell or expose them if the victim does not pay the ransom. Proposed ML techniques have been typically applied to the infection step, by using approaches to identify malware (that we describe in Sect. 6.3), and to data access denial step by using ML-based AD models that are trained to recognize specific process activities typical of ransomware. Therefore, for the training of these models, feature selection is critical.

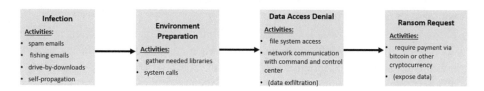

Fig. 6.2 Ransomware's workflow (based on Fig. 1 in [160])

Table 6.1 RWGuard's process features

Metric #	Metric name
1	Number of IRP_WRITE requests
2	Number of FastIO_WRITE requests
3	Number of IRP_READ requests
4	Number of FastIO_READ requests
5	Number of IRP_OPEN requests
6	Number of FastIO_OPEN requests
7	Number of IRP_CREATE requests
8	Number of FastIO_CREATE requests
9	Number of IRP_CLOSE requests
10	Number of FastIO_CLOSE requests
11	Number of temporary files created

An example of a comprehensive approach to AD for ransomware is the RWGuard system by Mehnaz et al. [155]. RWGuard uses two different sets of features, characterizing respectively processes behavior and changes to files. Concerning processes, RWGuard monitors 261 processes, including Explorer, Svchost, Chrome, GoogleUpdate, WinRAR, Taskhost, and Drpbx, and for each collects data on 11 metrics, each of which corresponds to a feature used by the ML-based detector for processes (see Table 6.1 for the list of these features). Key features are represented by the number of different types of fast I/O operations, as fast I/O is an activity typical of ransomware. The reason is that ransomware has to very quickly encrypt all the files to reduce the chance of being detected, especially when AD systems are very fast. For process AD, four different ML techniques were evaluated, namely Naive Bayes (using estimator classes), Logistic Regression (multinomial logistic regression model with a ridge estimator), Decision Tree, and Random Forest. The experimental results showed Random Forest to be the more accurate technique.

Concerning the file system, RWGuard monitors file changes to detect anomalous file changes; these anomalies are detected by using several indicators:

- *Similarity metric*: In comparison with a benign file change, e.g., modifying some of the existing text or adding some text, encryption would result in data that is very dissimilar with respect to the original data.
- *Entropy metric*: Entropy, as it relates to digital information, is the measurement of randomness in a given set of values (data), i.e., when computed over a file, it provides information about the randomness of data in the file. Therefore, a user's data file in plaintext form has low entropy, whereas its encrypted version would have high entropy.
- *File type change metric*: A file generally does not change its type over the course of its existence. However, it is common for a number of ransomware families to change the

file type after encryption. Therefore, whenever a file is written, RWGuard compares the file types before and after the write operation.

- *File size change metric*: Unlike file type change, file size change is a common event, e.g., adding a large text to a document. However, this metric along with other metrics can determine if the file changes are benign or malicious.

For detecting anomaly file changes, however, no ML model was used as just a simple custom algorithm was sufficient. In addition, RWGuard used an additional technique based on the notion of *decoy files*, that is, fictitious files that are not written by users/applications. As such, modifications to these files represent an indicator of a potential ransomware attack.

It is important to mention that the evaluation experiments of RWGuard on several ransomware families have shown that monitoring only the process activities or only the file changes is not sufficient for effective detection and results in both high false positives and high false negatives (e.g., the Cryptolocker ransomware encrypts files very slowly, which sometimes evades process monitoring), and that decoy files are effective. Therefore, combining process AD with file changes AD and decoy files represents a strong defense. Experiments have also shown that RWGuard is extremely fast in detecting ransomware, which is critical in order to quickly stop the attack and thus minimize the number of files that end up encrypted.

The design and experimental evaluation of RWGuard have given important insights into an effective AD. First, the selection of features is critical and requires insights about the activities of the specific malware—based on which one can determine the features to use. Second, effective AD must observe malware with respect to different points of view—such as processes and files as in the case of RWGuard. Different detectors may be used—some of which are based on ML techniques whereas others are not. Then the results of these various detectors must be properly combined to optimize the accuracy metrics of interest. Third, AD must be extremely fast as very early detection is critical to contain/block/respond to the attack.

6.3   Malware Detection and Classification

Malware, also referred to as malicious software, is still today one of the major threats. Attackers often install malware on the victim system by tricking the users into performing certain actions, such as clicking on some URL from which the malware is downloaded, or opening e-mail attachments, or downloading some files. To gain a foothold in the victim system, malware can also exploit vulnerabilities of the victim, such as bugs, open ports, and plugins. Once downloaded, the malware often spreads to other systems/networks.

To protect against malware, an important security function is to determine whether a software sample is malicious or not, and if the software sample is malicious, to determine its type. Therefore, malware detection techniques have been continuously enhanced to deal with the rapid evolution of malware [129, 222].

Many proposed approaches are based on ML-based classification techniques. Classification can be binary, that is, it categorizes the software samples into two classes, namely malicious and non-malicious (benign/normal), or multi-classification, that is, it classifies the sample according to different malware categories, such as backdoor, Trojan, and virus. Being able to correctly classify malware helps to gain knowledge about the behavior of the malware and thus to better defend against it. Malware detection is however challenging because of obfuscation and behavior modification techniques adopted by the malware authors.

In what follows, we first introduce the structure of the portable executable file format. We then briefly discuss existing malware analysis and detection techniques and their shortcomings. We then cover the issue of preparing malware code for training ML models. We then discuss the analysis of malware for different platforms, mainly focusing on the used features. We then discuss three different approaches to represent the entire malware code.

6.3.1   Portable Executable File Format

Malware analysis requires analyzing the malicious files by understanding the structure of the portable executable (PE) file format, as this format is often used by malware authors to create and manage their wrapped executable codes. The PE format encapsulates the necessary information, like dynamic link libraries, object codes, resource management, and API import and export, required by the operating system to manage the executable. The format consists of a series of headers and sections that carry information of various nature. The structure of the PE format for a "Cookrel2.exe" file (malware executable) is shown in Fig. 6.3.

The PE header contains some important contextual information related to file attributes that is useful for malware analysis. The timestamp information in the PE header helps in identifying the compilation time of the malware file. The optional header defines the entry point location of the PE file. The information about the entry point address is important for the malware analysts to begin with their analysis via reverse engineering. The optional header also carries information about the amount of the data initialized or uninitialized loaded into memory when a PE file is executed. The other essential information carried by the optional header is related to the size of the code, memory, and file alignments.

The section header maps the data and instructions for the loading process into memory and includes the permissions (read, write, and execute) that must be given to the code. The section header has special sections that are intended to handle different types of data and instructions:

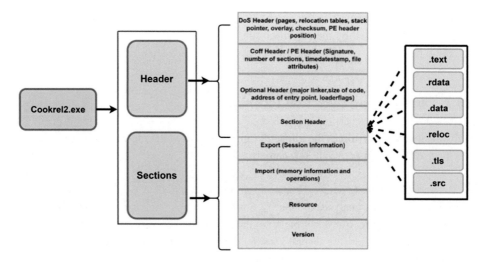

Fig. 6.3 PE file format

- .text: This section contains the executable code of the program. The x86 processor has an executable section header marked as .text.
- .rdata: It contains the initialized data which is only readable.
- .data: It is meant for the initialized data and also includes additional information such as button skins, cursor images, and so on.
- .idata: It contains the import address table (IAT), which includes the library calls and functions, and lists the dynamically linked libraries.
- .tls: This section provides storage for the program threads.
- .reloc: It contains the relocation information that allows the code to be executed without interruption even if it is moved during the execution.

The PE file structure also includes some optional fields like the export section, which is the mandatory information related to communication sessions, mainly secure socket layer/transport layer security (SSL/TLS) connection information. Malware makes use of SSL/TLS to hide network traffic from monitoring. The optional field also contains a checksum value to be used for checking the integrity of the file. The checksum value is calculated by the compiler when the execution file is created, therefore, any modification to the checksum invalidates the integrity of the file post compilation. The import section in the PE structure contains information regarding memory operations and thread libraries. The information about the resources accessed by the code is included in the resource section. The resource section also includes information related to dialog layouts, icons, manifest, and schemas. At the end of the structure, there is the version section, indicating the version number of the file with the creation date, and copyright information (if available). In most cases, the malware sets the checksum and version value to "0" to indicate that no checksum and version verification are required by the execution file, which is usually not the case for legitimate execution files.

6.3.2 Analysis and Detection Techniques

Malware analysis determines the impact, origin, and goal of the malware. Such an analysis may have different objectives [231], namely

- Malware detection: to detect whether a sample is malicious.
- Malware similarity analysis: to identify similarities among malware samples—for example to assess whether a new sample is similar to previously seen samples. As discussed by Ucci et al. [231], there are several versions of this objective, including variant detection, family detection, similarity detection, and difference detection.
- Malware category detection: to detect the specific type of malware with respect to a given malware taxonomy.

Malware analysis is generally categorized as static, analysis, or hybrid (see Fig. 6.4).

Static analysis focuses on the malware executable code and structure without executing it. Static analysis is useful for identifying malicious software samples and differentiating them from benign ones by generating fingerprints of the malicious samples. The malicious sample fingerprints help to uniquely identify the known malware. The features analyzed during static analysis include strings, file hashes, DLL imports, opcodes, API calls, and metadata associated with the PE file. Static analysis is however not very effective when the malware authors use dynamically downloaded data, malware packing, resource obfuscation, and anti-disassembly methods.

In contrast to static analysis, dynamic analysis monitors the behavior of the malware when is executing. It is typically carried out in a controlled virtual environment, like a

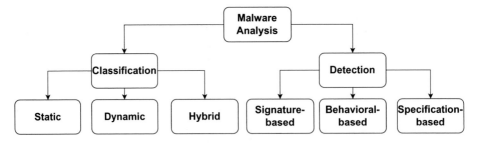

Fig. 6.4 Malware analysis techniques

sandbox. The actions executed by the malware are traced dynamically and extended to trace the traffic, API calls, registry edits, memory analysis, and modification to the system files.

Hybrid analysis combines static and dynamic analyses. Initially, the source code is analyzed without executing it and later the analysis is performed by executing the code in a controlled virtual environment. In other words, hybrid analysis is concerned with analyzing the signatures of the malware code and improving the analysis by combining it with the dynamic parameters.

With respect to detection, there are two main techniques, namely signature-based and behavior-based. Signature-based detection is based on fingerprints that uniquely characterize the different types of malware. It has low false positives, but it can be evaded by malware code modifications. Its main limitation is, however, in the case of unknown malware. The behavior-based detection monitors the malware actions dynamically and is effective for detecting unknown malware. However, it is prone to high number of false positives. In addition to signature-based and behavior-based techniques, there are specification-based techniques that impose some rules that specify the valid behavior, in order to distinguish it from malicious behaviors. The rules imposed are based on the features extracted and may be applicable for networks or systems depending on the detection site for malware. However, when obfuscation techniques are adopted by malware, one has to frequently update the rules which increase the complexity of the systems monitoring the malicious behavior. To address the shortcoming of the existing techniques, ML-based techniques have been proposed to enhance malware detection and classification and also to deal with large-sized data generated by commercial threat intelligence feeds.

6.3.3   Data Preparation and Labeling for ML-Based Malware Analysis

An important step for ML-based malware detection and classification is the training of the models, which then requires preprocessing malware datasets in order to generate proper feature vectors. Such a preprocessing involves recognizing malware patterns and data distributions, and identifying dependent and independent variables through exploratory data analysis (EDA) which involves visualizing the dataset through graphs mainly through scatter plots and histograms. To apply EDA, domain knowledge expertise is required for the interpretation of data. EDA is also helpful in identifying class imbalance issues and avoiding biased results from the models. For building the feature vector space, one typically applies feature extraction and feature selection techniques. We now illustrate such activity with a simple scenario showing how the feature vector is prepared for malware detection and classification.

Example

Suppose we have 50 malware files and 50 normal files in a dataset as in Table 6.2, and let's consider malware detection as a NLP problem. We consider the malware, the normal, and log files as documents containing codes, functions, connections, API calls, and other

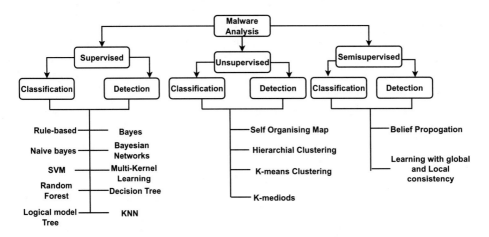

Fig. 6.5 ML-based techniques for malware classification and detection

Table 6.2 Malware feature extraction example scenario

# of samples	Analysis mode	File samples	Features
50 malware	Static	Bad_Rabbit.exe; cooker1.exe	API calls, Opcodes, strings
50 normal	Static	Calculator, Browser files	API calls, strings, opcodes
50 malware	Dynamic	Log file	Dll, registry keys, connections
50 normal	Dynamic	Log file	Dll, registry keys, connections

information. The log files are obtained after running the malware and normal files on an emulator (sandbox environment). The goal of the ML model is to find a pattern or difference between the malicious and normal code samples like finding the frequency of the calls to "dll23" expected to be higher in malware.

Using NLP to generate the feature vector space gives the following matrix:

$$
\begin{array}{c}
\\
s_1 \\
s_2 \\
s_3 \\
\\
s_{100}
\end{array}
\begin{array}{c}
mask\ dll23\ dll85\ \ldots\ urldown \\
\left[
\begin{array}{ccccc}
1 & 1 & 0 & \cdots & 1 \\
1 & 0 & 1 & \cdots & 0 \\
1 & 1 & 1 & \cdots & 1 \\
\vdots & \vdots & \ddots & & \vdots \\
0 & 0 & 1 & \cdots & 1
\end{array}
\right]
\end{array}
\begin{array}{c}
target \\
\left[
\begin{array}{c}
M_1 \\
N_2 \\
M_3 \\
\vdots \\
N_{100}
\end{array}
\right]
\end{array}
\qquad (6.1)
$$

The row labels indicate whether the file is malware or normal, and the column label indicates whether or not the file contains functions, dlls, connections, and memory information that contribute to relevant features. The target variable denotes whether the file is malicious "M_n" or normal "N_n" based on the values in the feature space. A "1" value denotes the presence of a feature in the sample file, like "dll23" which is a feature present in the malware file and not in the normal one. Similarly, a "0" value denotes the absence of a feature. The feature vector created in this way is best suited for supervised classification models as it contains well-defined labels.

In the case of high-dimensional feature space, PCA and t-distributed stochastic neighbor embedding (t-SNE) can be applied to reduce the data dimensionality and visualize the distribution of the data for selecting the important features from the dataset.

The feature selection for malware can be carried out using filter-based or wrapper-based methods. The filter-based method focuses on the intrinsic characteristics of the features like information gain, correlation, consistency, and distance. On the other hand, the wrapper-based method uses learning algorithms like Support Vector Machine (SVM), Neural Networks (NN), and Random Forest (RF) to evaluate the importance of a feature for classification.

The model is then trained on the preprocessed dataset and, as customary when training ML models, then validated and tested (see Fig. 6.6). We notice however that in some cases the validation phase is omitted. Once tested, the model is then deployed.

A challenge for the use of supervised ML models is the manual effort required for labeling the malware samples. To address this challenge, Rizvi et al. [190] propose a DL-based unsupervised classification technique that uses the feature attention block for static analysis. The main idea is to train the DL framework on pseudo-labels and use the k-means clustering algorithm which partitions the samples into clusters of benign samples and malicious samples. The use of the attention mechanism is to prioritize the PE features (see Table 6.3), which have a higher impact than other features on malware detection.

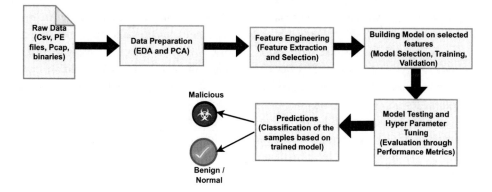

Fig. 6.6 ML workflow for malware detection

Table 6.3 Essential features for malware analysis

Features	Description	Feature for ML-based analysis
Byte sequences	Specifying the number of times, a sequence and combination of n-bytes have occurred	n-gram
Opcodes	Machine-level interaction of PE	Opcode sequence
API and system calls	List of calls executed and invoked by the PE file	Execution traces
Network activity	Interaction of PE with the network regarding C and C server	Protocols, DNS interactions, HTTP requests, TCP/UDP ports
File system	File operations done by malware to gain persistence	Number of files created, modified, encrypted, deleted
CPU registers	Registers or hidden registers used by the PE	FLAG register values
PE file characteristics	Information about the PE file sections	Imports, compilers, symbols
Strings	Strings contained in PE file carrying information regarding	Signatures, file names, code fragments
Control Flow Graphs (CFGs)	Information about the programs structure	Functions, control relationships, code sequence

6.3.4   Malware Detection and Analysis: Features for Specific Platforms

In what follows, we discuss malware features used by ML-based models for analyzing malware on different platforms, namely Android-, Windows-, and Linux-based systems.

- **Android Malware Analysis.** The main concern for Android security arises from Android ability of easily invoking third-party code, which may include malicious codes, and its coarse-grained permissions. To enhance security, Android's latest versions provide several security capabilities, such as lock mode, access control for cameras and microphones, encrypted backup, file-based encryption, Google Play protection mechanism, and access control for sensitive information. However, such security capabilities are not always able to protect against malware, as there are apps that look legitimate to users and ask for permission from users for downloading themselves like other benign applications. The permissions granted by users then allow the attackers to access the ".apk" files and exploit the permissions to install themselves and access device information. Therefore, permission-based controls fail to block malware. To enhance security, ML-based models have been proposed that rely on extracting meaningful features from the APK files,

Table 6.4 Features for malware detection in Android

Features	Description
Permissions used	Requested permissions by apps
Application components	Include Android components like service, activity, broadcast receiver, and content provider
Filtered intents	Messages passed between Android components
API calls	Calls to invoke core functionalities
Network addresses	URLs and IP addresses used by the application
Opcodes	Operation codes and their sequences in the application
Native code	Function calls to external libraries
Hardware components	USB, Camera, GPS accessed by the application
API Call Graph (CG)	API calls between modules depicting nodes as APIs and edges as links between the nodes
Control Flow Graphs (CFGs)	Represent the code blocks and transitions between the blocks
Data Flow Graphs (DFG)	Information flow between the program entities

based on which one can detect malware. The features to be used can be selected by carrying out a feature validity analysis using chi-square tests, correlation analysis, conditional entropy, information gain, and posterior probability. The features typically used for Android malware analysis are listed in Table 6.4 [207].

The "manifest.xml" file is a binary file that consists of all the permission requests made by the apps. This file is converted to a text file before carrying out the malware analysis. The feature vector is created by selecting the important features and assigning the binary values ("1" if the feature is selected and "0" if the feature is not selected). Android applications can be classified as malicious or legitimate based on multiple semantic features as well [20]. The semantic features are useful for identifying behavioral semantics, like n-gram sequences, source or sink methods, dynamic code methods, and sensitive permissions selected based on the activation events, context semantics, and data flow environment indicating whether the flow is malicious or not [247]. A particle swarm optimization-based technique is used by Azat et al. [20] to extract the features from the publicly available Android malware dataset CICAndMal2017 [139] containing samples of both benign and malicious applications with a dimensionality of 919. Samples are classified using NNs with 2 hidden layers all densely connected. The model is trained using the stochastic gradient descent (SGD) algorithm with backpropagation; the feature vector is fed forward through the layers. The experiments show that the model achieves a true positive rate of 83.4% and an F-measure of 82.5%.

- **Windows-based Malware.** For detection purposes, API-based features are important for understanding the program's behavior. Most ML-based approaches use API's statistical features, like malicious sequence patterns or calculating the API call frequencies,

Table 6.5 Features and extraction techniques for detecting Windows malware

Feature extraction technique	Features extracted
Signature extraction	Byte sequence hash code, file Hash codes
DLL function call extraction	List of DLLs used, function calls, number of function calls
Binary sequence extraction	Hexadecimal code of each file, Binary sequence of n-gram, mapping each byte to feature vector
Assembly sequence extraction	Assembly sequence, opcode sequence, n-grams
PE file Header fields extraction	DLL bindings, API import, exports, threads, sources, file size, packer info, section alignment, sizeOfimage, COFF file header
Machine activity metrics	CPU usage, Context switches, SWAP use, Number of Processes running, sent and received bytes, network information, memory information
Entropy signals extractions	Entropy of code chunks, signal values, file as entropy stream

in order to distinguish malicious behavior from a normal program behavior. However, such approaches can be easily evaded by the attackers by shuffling the API sequence or inserting irrelevant calls. Therefore, only using API sequences and patterns is insufficient for effective detection. To address such an issue, an important source of information is represented by the actual API calls. API calls carry different functionalities, which can be determined based on their functions and function parameters. The approach by Rabadi and Teo [184] thus also uses information about the API calls, functions, and function parameters from the collected execution traces to analyze the program's behavior and select the unique features by performing a similarity analysis between the API functions using a Markov chain. ML models are trained to learn from the extracted features, using a 5-cross-validation approach, to distinguish malicious API calls from normal ones.

However, as even using API calls may not be sufficient for an accurate malware detection, additional features can be extracted, using different techniques. Examples of those features include DLL function calls, binary sequences, assembly sequences, PE file header fields, machine activity metrics, and entropy signals [73]. It is important to notice that different features often require different extraction methods. Table 6.5 lists the features that can be used for malware detection and classification, and the corresponding extraction method.

- **Linux-based Malware**. Linux has an effective built-in user privilege model, built-in kernel security defenses, and kernel lockdown mechanism. However, despite those security mechanisms, Linux is also prone to malware attacks. Malware typically infects the running processes by modifying the environment variables, which allows access to libraries

Table 6.6 Features for Linux malware analysis

Static features	Dynamic features
Text Strings	Unique system calls
Cyclomatic complexity	Ioctl
Basic blocks	Renamed processes
Machine code	Number of processes created
Library functions	Check user and group identifier
Entropy	Information of opened files

and functions. After gaining access, the malware typically collects commands and login passwords in order to give remote access to the attackers.

To study and analyze the behavior of Linux malware Carrillo-Mondejar et al. [38] applied different ML algorithms to mine information from ELF headers and train the model with static, dynamic, and hybrid features. The features used for training the model are listed in Table 6.6. The labeled and unlabeled samples were distributed among different architectures: ARM32, MIPS I, AMD x86-64, Intel 80386, and PowerPC.

The approach by Carrillo-Mondejar et al. [38] recognizes unknown threats from unlabeled malware data samples by clustering the samples based on features and similarity index. Different learning algorithms already trained on labeled datasets were used to identify known malware patterns in the unlabeled data. The study was further extended to identify the unknown clusters for unlabeled datasets by using identified known patterns and distance metric-based clusters. The performance of different classifiers was then evaluated on unlabeled malware datasets.

An interesting observation is made by Sentanoe et al. [200] regarding the sequence of system calls. The sequence indicates the start of the SSH session, which can be used for recreating the session and reproducing malicious files. However, if the sequence calls are of variable length, hypergram sequences [153] can be used. The hypergram sequences represent the n-gram sequences in a k-dimensional hyperspace, where "k" denotes the unique system call. The API calls can be combined with knowledge-based meta-features to improve the classification accuracy of the model [253].

6.3.5   Malware Representation

Even though feature-based detection of malware is generally effective, features alone are not sufficient to always fully characterize malware. Also, models based on features alone may be bypassed by malware using evasion techniques, such as obfuscation and alterations to the sequences of calls. Characterizations by which static and dynamic features are organized according to a representation are critical to allow ML tools to better analyze malware. Important information to be captured by such a characterization includes

- Order of actions performed by malware.
- Actions performed on intended file or object.
- Program control flow and data flow.

To address the need for such comprehensive characterizations, three different malware representations have been proposed that we discuss in what follows.

Graphs

A graph representation can be based on system calls, function calls, API calls, and control flows. System call graphs are used to represent the system call traces of a program having processes as vertices and system call invocations as edges. In function calls graphs, vertices represent functions and associated attributes and edges represent caller-callee relationship. Function call-based graphs are less prone to instruction-level obfuscation techniques adopted by malware for evading anti-virus software. API call graphs are used to represent the API calls and their associated functions. Control flow graphs (CFGs) represent the control flow structure of the program where each node corresponds to a basic block of the program and edges represent the control flows between blocks. In order to be used by ML models, such graphs are typically transformed into vector forms through the use of embedding techniques (see Sect. 2.8). A challenge in the use of graph representations is to find similarities between the samples based on the structural similarities of the graphs as the computation of the similarities among graphs can be computationally prohibitive.

To address such a challenge, graph neural networks (GNN) have been used. The approach by Yan et al. [249] uses deep graph convolutional neural networks (DGCNN) to handle variable-size graphs through fixed-sized tensors employing the Graph2vec model for feature extraction. Graph2vec—a model similar to the Doc2vec model [157]— is used to convert graphs into vector representations. Graph2vec uses skip-grams to cluster similar graphs based on the code sequences, structures, and attributes of the neighbors.

The limitation of the Graph2vec technique is that it suffers from an unnecessary neighborhood expansion, increasing the space complexity. Possible solutions include partitioning the graphs into subgraphs, applying the Node2vec model for generating a node-to-vector representation, and restricting the search to the subgraphs of the node. The Node2Vec model [40] captures the features of each node from the graph and maps the nodes of the graph to an embedding space. The embedding space has dimensions lower than the number of nodes in the original graph.

Abstract Syntax Trees

Abstract syntax trees (ASTs) are used for the intermediate representation of source code. A code can be represented using an AST by traversing the code line by line and gathering the class names, functions, and other objects of the code and representing them as nodes. The edges are represented by the sequence of instructions, including branching instructions (i.e., function calls) and iterations.

Before the training phase, an AST is divided into subtrees based on the functionality carried by the respective code sections represented in the tree to determine the malware obfuscation transformations. The subtrees are then parsed to build an embedding matrix containing nodes and node types (e.g., object expression, variable declaration) that describe the subtrees. Figure 6.7 shows an example of a malicious code snippet (on the left-hand side) for string encryption. Then the code is parsed (i.e., tokenized) to capture various information such as keywords, function calls, and variable declarations. Finally, the AST is generated based on the tokenization (on the right-hand side) step.

Rusak et al. [196] used the Doc2Vec model to extract the necessary features from AST nodes and edges obtained by code tokenization. Then, they trained a DL classifier on the fixed-length feature vectors and generated a similarity matrix that measures the similarity between normal subtrees and malicious subtrees [196]. When the model is applied to a malicious code not seen during model training, the AST representation of the code is generated, and the feature vectors are extracted using the Doc2vec model. Then the model classifies the code as normal or malicious. Ndichu et al. [167] have found that Doc2vec model

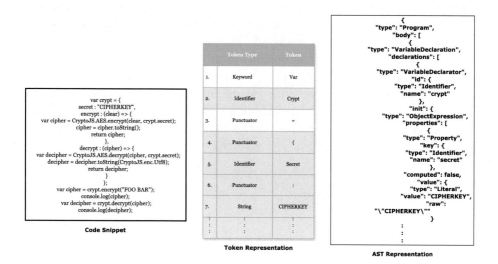

Fig. 6.7 AST representation of malicious code snippet for string encryption

hyperparameters, such as the minimum frequency of words, the distance between the words, and the vector size (i.e., feature vector dimensionality), can be tuned to improve the classification accuracy.

Images

Malware representation using images was inspired by progress in ML learning techniques for image analysis. The approach is based on the observation that a malware executable can be represented as a binary string of zeros and ones. Such a vector can then be organized into a matrix and viewed as an image [166]. In their experiments, Natraj et al. observed significant visual similarities in image textures for malware samples belonging to the same family. Therefore, based on such insights, malware can be classified, according to families, by using ML models for image classification. More recent approaches based on the same ideas are by Dai et al. [60], which use grayscale images or red-green-blue (RGB) representation and pre-trained networks like RESNeT, and by Huang et al. [104], which use images for representing both static and dynamic analysis results.

6.4   Research Directions

Detection is an important cybersecurity function with several interesting research directions. One important direction is to enhance models to reduce the number of false positives, especially for ML models deployed in dynamically changing environments. Also enhancing the robustness of the models against adversarial attacks aiming to evade the models is critical. Approaches to ensure robustness have been developed mainly for image applications, and they need to be revised/extended for data specific to cybersecurity functions, such as tabular data. Also concerning malware classification and analysis one interesting direction would be to use multiple representations, instead than just one. The reason is that different representations may provide different complementary information.

Attack Management

7

Even well-secured systems can be breached. There are several reasons for this, including the complexity and dynamic nature of today's systems, insider threats, zero-day vulnerabilities, and human errors. Therefore, as we have discussed in Chap. 6, monitoring and anomaly detection activities are critical. However, one important task in the security life cycle is to take proper actions, based on the results of those activities. The goals of those actions include mitigating, containing, and recovering from the attack, while at the same time trying to maintain the system operational as much as possible. In this respect, it is important to mention that resilience to attacks is today increasingly critical as many systems have stringent continuity requirements.

Also, analyses, referred to as digital forensics or computer forensics, of attacks are critical in order to determine the steps of the attacks and the vectors used in each step, so as to eliminate vulnerabilities, exploited by the attacks, and enhance defensive mechanisms.

To date, the area of attack mitigation, containment, and recovery has not been widely explored, as the actual approaches to use depend on the application domain and the security and resilience requirements. It is however an area for which the application of ML techniques, and especially RL techniques is promising and a few interesting approaches have been recently proposed. By contrast, the area of digital forensics has been widely investigated over the past [126, 195], and many ML-based approaches have been proposed [201].

In what follows, we first present an approach leveraging RL for attack mitigation in networks; this approach provides an example of actions that can be taken to mitigate a potential attack, while at the same time ensuring that the system is able to continue working with good performance. We then present an initial approach, based on the notion kill chain, to enhance ML-based defenses by leveraging the ML attention technique. We then briefly discuss ML-based digital forensics.

© The Author(s), under exclusive license to Springer Nature Switzerland AG 2023   105
E. Bertino et al., *Machine Learning Techniques for Cybersecurity*, Synthesis Lectures
on Information Security, Privacy, and Trust,
https://doi.org/10.1007/978-3-031-28259-1_7

7.1   Attack Mitigation

In general when dealing with attack mitigation, one has to take into account that early mitigation is critical to reducing the effects of the attack, but at the same time an anomaly detector may need more data before having high confidence that an attack is actually ongoing. Therefore, depending on the specific system/domain, attack mitigation strategies have to take actions based on the trade-off.

An example of such an approach is represented by Jarvis-SDN, a RL-based system, designed to mitigate attacks in software-defined networks (SDNs) [162]. The goal of Jarvis-SDN is to optimize rate control for network flows by detecting and then blocking malicious flows. It is important to notice that reducing the number of false positives is critical. A large number of false positives would result in blocking a a large number of benign flows, thus resulting in very poor rate control.

RL-based approaches for network control typically focus on optimizing a single functionality, which makes them not suitable for applications in real networks. For example, learning a policy which maximizes the throughput of the network (functionality 1: optimal routing) may have the cost of unfair bandwidth consumption by a set of users (functionality 2: per user bandwidth fairness). More challenging is the case of security policies. For example, learning a policy which maximizes the throughput of the network (functionality 1: optimal rate control for QoS) can unknowingly facilitate the propagation of a high throughput DoS attack. To address such a challenge, Jarvis-SDN uses a reward defined as a weighted combination of individual functionality performance metrics, including a metric for security behavior of the system.

A challenge when using RL for security is the definition of security metrics as it is not obvious how to define performance metrics for security (see [180] for a discussion on security metrics). To address such a challenge, the approach adopted in Jarvis-SDN for quantifying security is to measure the ability to detect known attacks. The approach requires to first build a set of "attack signatures" from packet captures of previously seen attacks. Those signatures are then used by the RL agent to determine a security quality value for the network state depending on the perceived threat a network flow has on the current and near-future states of the network. Then once the set of signatures is built, a RL-DQN-based IDS is trained. Once trained, the IDS analyzes network flows and reports at each given interval the security quality value to the RL agent controlling the rate.

The reward function of Jarvis-SDN is thus defined as follows:

$$R(S_t, A_t) = (1 - \delta)[f_1 F_1(S_t, A_t) + f_2 F_2(S_t, A_t)] + \delta D(S_t, A_t)$$

where S_T denotes the state at time t, and A_t denotes the action taken at time t; δ, f_1, and f_2 are hyperparameters to incorporate security metrics, rate control, and user fairness weights in the objective function.

The normalized functionality and security metrics used in the reward function are defined as follows. We observe that the rewards are expressed as expectations over all the flows in the network.

- *Rate Control*: It gives a positive reward proportional to the amount of traffic allowed when the current load is less than the threshold and a large negative reward for overloading the server. The positive rewards are proportional to the amount of traffic let through, which encourages higher throughput.
- *User Fairness*: It gives a positive reward proportional to the amount of traffic allowed when the service-level agreement is upheld and a large negative otherwise.
- *Security*: It is the quality security value generated by RL-DQN IDS built based on the set of attack signatures. It gives a positive reward when the flow matches a benign flow and a negative reward when it matches a malicious flow.

Another important aspect when using a RL-based approach is related to the space of actions that the RL agent can take. In Jarvis-SDN, at each interval the agent can take 11 possible actions {0, 10, 20, ...,100} corresponding to percentages of packets to drop while allowing the rest to flow. 0 corresponds to allowing all traffic to pass and 100 corresponds to dropping all traffic.

Jarvis-SDN has been evaluated via simulation. An interesting result from the experiments is that Jarvis-SDN is able to detect malicious flows while at the same time minimizing the number of false positives. The reason is that when Jarvis-SDN observes some initial packets of a flow, and these packets are classified as malicious by the IDS, Jarvis-SDN just slows down the flow and waits to analyze more packets. As a result, a few malicious packets can still be transmitted but at a lower rate, which makes most attacks ineffective. However, on the other hand, benign flows whose initial packets may be classified as malicious are not blocked, even though some of their packets may be transmitted at a lower rate.

7.2 Defense Enhancement

As we discussed in Chap. 6, ML-based detection techniques are an important defense mechanism. However, their accuracy can be undermined for various reasons, namely model evasion techniques and poor training. Model evasion techniques allow an attacker to craft malicious input that is not detected as malicious by the detector (see Sect. 9.4). Poor training is often due to the lack of training data of good quality or changes in the protected systems (see the discussion on robustness in Sect. 9.3.3). Therefore, a key requirement is to develop techniques to enhance the accuracy of detectors based on analyses of data collected during the attack.

A novel approach addressing such a requirement has been recently proposed by Lee et al. [142] for the case of IoT systems. The approach is based on several insights:

- Attacks have several steps; these steps are often referred to as kill chain. Although the number and the name of steps vary, these kill chains commonly break down an attack into the following five steps:
  - **Reconnaissance**: the attacker collects information about the target. The attacker may perform social engineering or port scanning.
  - **Infection**: the attacker exploits vulnerabilities of the target to take over and install malware binaries required to launch attacks. The attacker may launch dictionary attacks or zero-day attacks for this purpose.
  - **Lateral movement**: once the attacker has access to the target, the attacker may move laterally to other devices to gain more leverage. The attacker may perform scanning activities or propagate malware in the internal domain.
  - **Obfuscation**: the attacker hides its tracks. This step may involve laying false trails and clearing logs.
  - **Action:** the attacker launches the attack. For example, in botnets, the attacker directs bots to perform a DDoS attack such as UDP flooding.
- A robust defense is based on allocating several detectors based on the kill chain. Each such detector is specialized in detecting anomalies for a given attack step. Such an approach raises the bar for the attacker in that the attacker would have to evade all the detectors in order to succeed (see also discussion in Sect. 9.4).
- Today, systems and networks provide logging facilities by which one can collect data to analyze security-relevant events, such as network packets.
- If a detector D misses to flag a malicious event part of an attack, it is likely that some of the other detectors will detect other malicious events related to the same attack. Then based on the log, by using correlation techniques, one may identify the activity missed by detector D and use the data in the log to re-train the detection model of D.

Based on those insights, Lee et al. [142] designed IoTEDef, which detects attacks based on analysis of network logs and subsequently retrains detectors. IoTEDef is based on a kill chain consisting of three steps, namely reconnaissance, infection, and action, and thus includes three detectors—one for each such step. Its goal is to enhance the accuracy of the detectors associated with the early attack steps, that is, reconnaissance and infection. The reason is that in order to acquire a foothold in a target system, an attacker typically attempts to evade the detection systems by performing stealthy attacks (e.g., stealthy, distributed SSH brute-forcing) or exploiting unknown device vulnerabilities (i.e., zero-day attacks). To detect such attacks, the detectors may classify all the suspicious or unknown patterns as anomalies, but it would result in a high number of false positives.

Upon detecting anomalies related to a later stage of the kill chain, IoTEDef backward-traverses the log of the events and analyzes these events to identify infection events. Then based on the identified events, IoTEDef improves the performance of its infection detector. Essentially, IoTEDef gives up precise detection of unknown patterns of an early stage attack when it faces an unknown attack for the first time. Instead, after IoTEDef recognizes that there has been an early stage attack through known patterns of the later stage attack, IoT-EDef identifies the corresponding early-stage attack patterns and makes its early detector learn the patterns. Then, IoTEDef can detect such an early-stage attack with high precision later on.

A critical issue is the correlation between adversarial and separate events in two different steps (e.g., UDP flooding in the action step and dictionary attack packets in the infection step), where the interval between them can be long. Addressing such a challenge requires mapping diverse networking patterns onto kill chain steps and backtracking from later events to earlier events. The approach designed for IoTEDef is based on the observation that similar challenges are also present in the area of language translation, and many techniques have been developed to address these challenges [43]. Therefore, the approach taken for IoTEDef is to model the event correlation problem as a language translation problem by introducing a novel probability-based embedding to encode past events into steps that the events belong to and an attention-based infection identification algorithm to correlate the encoded events with long-term dependencies in different steps. Such an approach allows one to identify infection events that lead to action events. The identified infection events are then used to improve the performance of the infection detector. Extensive experiments show that the attention mechanism is effective in enhancing the accuracy of the early detectors also with respect to other baseline approaches.

7.3 Digital Forensics

Digital forensics is a very rich area, which is articulated in many subareas, including operating system forensics, disk and file system forensics, live memory forensics, web forensics, email forensics, network forensics, and multimedia forensics. Many forensics techniques and tools have been also developed over the years. These tools are usually tailored for specific forensic tasks, such as attribution, alibi and statement, intent, provenance of artifact and data, and document authentication, and we refer the reader to [126] for a recent comprehensive survey of forensics tools and techniques.

Because of the relevance of forensics, also from a legal point of view, it is not surprising that forensics is an important application area for ML in the context of security. As discussed by Shahzad et al. [201], ML techniques can reduce the time required for finding evidence from vast amounts of data generated from different sources. They can also augment evidence for nonstructured data by building interlinkages and identifying otherwise hidden patterns, and identify malicious artifacts via prediction techniques. Many ML-based approaches have been proposed, which use a variety of techniques including DL, DM, and SVM, and we refer the reader to [173, 201] for discussions and surveys. More advanced ML techniques have been recently used for digital forensics. For instance, Alsaheel et al. use NLP techniques such as lemmatization and LSTM [7], and Ede et al. use transformers for digital forensics [232]. In the following, we look into the details of these advanced approaches.

7.3.1   NLP-Based Attack Analysis

Advanced Persistent Threats (APT) involve multiple attack steps over a long time period, and their investigation requires the analysis of myriads of logs to identify these steps. To automate this highly manual task, Al Saheel et al. [7] have developed ATLAS, a framework that constructs an end-to-end attack reconstruction from off-the-shelf audit logs. The main idea is that different attacks may share similar abstract attack strategies, regardless of the vulnerabilities exploited and the payloads curated. ATLAS leverages a combination of causality analysis, NLP, and ML techniques to build a sequence-based model, which establishes key patterns of attack and non-attack behaviors from a causal graph. At inference time, given a threat alert event, an attack symptom node in a causal graph is identified. ATLAST then constructs a set of candidate sequences associated with the symptom node, uses the sequence-based model to identify nodes in a sequence that contribute to the attack, and unifies the identified attack nodes to construct an attack story. For NLP, ATLAS uses word embedding to map the lemmatized sequences to a vector of real numbers, which captures the context of an entity in a sequence and its relation with other entities. This sequence-based representation then is fitted into the training of a LSTM model. The overview of the approach is shown in Fig. 7.1.

7.3.2   Transformer-Based Contextual Analysis of Security Events

Security monitoring systems detect potentially malicious activities in IT infrastructure, by looking for known signatures or anomalous behaviors. Security operators investigate these events to determine whether they pose a threat to their organization. In most cases, single events are not enough to determine whether certain activity is indeed malicious. Correlating multiple events is a hard manual task. Therefore, van Ede et al. [232] have designed DEEPCASE, a system that leverages the context around events to determine which events require further inspection. Figure 7.2 gives an overview of the approach. The core idea is to analyze, besides the security event itself, the preceding security events. Those preceding events represent the context in which the event is triggered. An event in combination with its context is called a sequence. DEEPCASE leverages a deep learning model with trans-

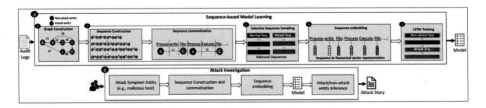

Fig. 7.1 Overview of ATLAS [7]

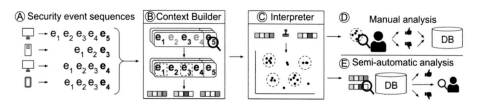

Fig. 7.2 Overview of DEEPCASE [232]

formers designed to expose complex relations between both new and previously observed event sequences. These sequences are subsequently grouped based on a similar function, providing concrete information about why various event sequences are treated as the same.

7.3.3   GNN-Based Memory Forensic Analysis

Memory forensic analysis can provide live digital evidence of attack footprints by analyzing the memory snapshot or dump of a running machine. For instance, by identifying the memory objects that hold the process information, analysts can know all the running processes when the attack happens.

However, memory forensic analysis is challenging for several reasons. First, memory analysis requires a large amount of manual effort to understand the target system and write rules to guide the analysis process. Second, the manually written searching rules can be easily bypassed by privileged attackers. For instance, once the attacker has gained control over the victim machine, the attacker could use techniques such as direct kernel object manipulation (DKOM) to modify the kernel object values to avoid being caught by memory forensic analysis tools. Finally, analyzing the memory dump/snapshot of the machine is inevitably expensive, especially when the analyst often has to analyze snapshots taken on several machines at different times.

To address such challenges, Wei et al. [244] propose the use of ML for memory forensic analysis (see Fig. 7.3). To enable the use of existing ML techniques for such analysis, they propose a graph-based representation of the memory dump and a graph neural network (GNN) architecture that performs classification tasks on graphs.

The memory dump, in practice, is represented as a sequence of raw bytes. However, this representation makes it hard for models to learn because it does not explicitly provide several vital information such as the relationship between different bytes. For instance, if the value of a memory byte is the address of another memory byte, it is likely that the two bytes have points-to relationship. To address such an issue, a more effective representation of the memory dump is adopted in DeepMem by representing the dump in a graph consisting of nodes and edges. Each node in the graph is a segment of memory. Nodes are connected by edges of different types. Different types indicate different relationships between nodes such as adjacency and points-to relations.

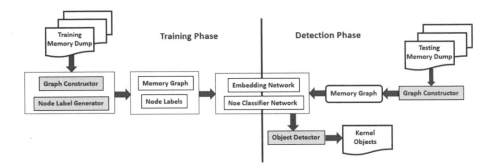

Fig. 7.3 Overview of DeepMem (redrawn from Fig. 1 in [244])

Given a memory dump, DeepMem generates a graph based on it and feeds it to the model. The model consists of two major components. First, a GNN component executes messaging passing on the whole graph to learn a vector representation (referred to as embedding) of each node. Then, the second component—a multi-layer fully connected neural network—takes the embedding as input and performs multiclass classification for each node. At the end, each node (each memory segment) in the graph is classified into a class that represents a certain type of data structure. Then further analysis such as process information extraction can be easily carried out based on the classification results.

DeepMem was evaluated over 400 1GB memory dumps on the Windows 7 operating system. Wei et al. even enabled kernel address randomization to increase the challenge of the memory forensic analysis. The results of the model performance are satisfactory. The precision, recall, and F1 values of the model are higher than 99%, showing the model's high accuracy in identifying different memory objects in the dump. In another experiment, they evaluated if the model can still maintain the accuracy when the attacker performs certain malicious modifications to the kernel objects. However, the model still achieves 100% precision and 99.77% recall.

7.3.4 An Explanation Method for GNNs Models

Although GNNs can achieve high performance on many real-world tasks including security applications such as code vulnerability detection, users cannot easily understand the reasons behind the model prediction by the GNN (e.g., why the GNN would predict a code block as vulnerable). Thus, a tool that can understand the process according to which the model makes the precision and then provides the explanation to the user is critical.

However, explaining GNN models is challenging. First, due to the models and input complexity (e.g., a GNN model might have 1 million parameters), it is hard to both comprehensively and accurately identify all the important factors that contribute to the model prediction. Failing to do either would hinder the real-world usability of the tool. Second,

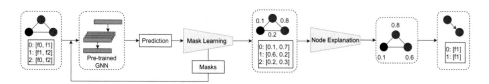

Fig. 7.4 Overview of ILLUMINATI (redrawn from Fig. 2 in [98])

an effective method needs to support various GNN architectures. Recently, many different GNN architectures have been proposed, and they are quite different from the basic GNN design. An explanation method that only works for specific GNNs will not be useful in practice.

To address such challenges, Hayou et al. [98] have proposed a new explanation approach and implemented ILLUMINATI—a prototype explanation tool (see Fig. 7.4). ILLUMINATI uses a two-stage workflow to explain a GNN model. First, it takes the trained GNN model and the model input as ILLUMINATI's input. It performs masking learning on the model, during which different edges or node features are removed from the input graph and analyzed in order to determine if the GNN model can still provide an accurate prediction. Intuitively, an important edge—if removed from the original input graph—may largely affect the model prediction.

Second, ILLUMINATI computes a score for each node in the input graph, indicating the importance of this node when making the prediction. The important score is computed based on the important edges and node features identified in the first stage. Once the score is computed for each node, ILLUMINATI extracts a subgraph from the input graph and reports this subgraph to the user, which consists of crucial nodes that contribute to the model prediction.

To evaluate ILLUMINATI, Hayou et al. choose the Essentialness Percentage (EP) metric, which is the percentage of subgraphs that retain the original model prediction. In general, an explanation method that has higher EP is more accurate in providing the user with useful information. ILLUMINATI is compared against existing explanation methods such as PGExplainer, and it has over 10% higher accuracy than them.

7.4   Research Directions

The area of ML-based attack management has many interesting research directions. The first is the prediction of possible next steps of attacks. As we have already mentioned, attacks are in most cases multi-steps. Therefore, based on some initial knowledge—based for example on the detection of an anomalous event—one could determine whether the event is related to an attack, and if this is the case one could determine the potential goal of the attack and then the next possible steps. Identifying the potential next steps of the attack would then allow the defender to execute actions to block, contain, and possibly respond to the attack. A challenge is, as in other cases, represented by the training of suitable ML models, which for example would require large sets of attack graphs, and also domain knowledge, such as vulnerability databases and details on the system under attack. A second research direction is the forensics of ML models and algorithms, and ML-based systems. There are many challenges, such as the difficulty of explainability of ML (see discussion in Sect. 9.3.1), which makes it very difficult to pinpoint specific reasons for incorrect/malicious decisions/previsions by these models. Reasons can include wrong parameter settings and malicious training data. Therefore, suitable forensics processes would require fine-grained and comprehensive logging capabilities able to capture all activities involved in training and using these models. Also as ML models and algorithms will be increasingly able to autonomous evolve, forensics processes will require detailed tracking and recording of this evolution, including contextual information.

Case Studies

8

In this chapter, we briefly discuss three case studies with the goal of highlighting, in addition to the conventional defenses, the ML-based defenses that could have prevented/mitigated the different steps of the attacks. Notice that in some cases, conventional defenses and ML-based defenses would be alternatives to each other; in other cases, they should be both deployed for enhanced security.

These case studies cover attacks with different goals; also the covered attacks differ with respect to some of the attack steps. The first case study covers the Target data breach, by which attackers exfiltrated large datasets of customer data and credit card data. The second case study covers The recent SolarWinds attack, by which attackers were able to inject malicious code inside a patch, which was then distributed to a large number of customers. The SolarWinds attack represents an example of supply-chain attacks, in which instead of directly attacking the victims, the attackers target a supplier of some software and inject malicious code into this software. The malicious code is then unknowingly distributed by the supplier to all its customers. The third case study covers the WannaCry ransomware, as this is a representative of today's most common attacks; most ransomware attacks have targeted healthcare and local governmental organizations, and also corporations. Recently however, they have targeted critical infrastructures, and in these cases the consequences can have disastrous consequences.

8.1   The Target Data Breach

The Target data breach is considered one of the biggest security attacks in history. The attack happened in 2013 and affected 41 million customers, causing Target to pay $18.5 million multistate settlement. This attack is a clear example of the worst outcome of a data breach,

© The Author(s), under exclusive license to Springer Nature Switzerland AG 2023   115
E. Bertino et al., *Machine Learning Techniques for Cybersecurity*, Synthesis Lectures
on Information Security, Privacy, and Trust,
https://doi.org/10.1007/978-3-031-28259-1_8

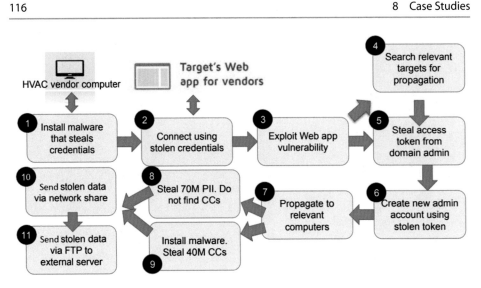

Fig. 8.1 Overview of the target data breach

which is customers losing confidence in the security of the system and refraining from the service (purchasing from Target) even after the security issue was resolved. Similar to many breaches, the attackers did not go directly through Target's system, but through a third-party vendor. Companies should make sure that their suppliers follow good security practices. In this section, we give an overview of the steps of the attack (see Fig. 8.1) and discuss potential ML-based techniques for prevention and/or detection and mitigation. For additional details on the attack, we refer the reader to the report by Aorato Labs [17].

Step 1: Installing malware to steal credentials from Target's vendor

The attackers first infected the system of Target's HCAV contractor with a malware known as Citadel, which can gather web application credentials from infected machines. The malware infected one of the contractor's machines through phishing emails.

Non-ML Defenses: Effective email spam filtering mechanisms are critical in filtering out illegitimate emails. Also, user education about fishing and spear phishing attacks is important.
ML Defenses: Adoption of advanced ML-based solutions combining NLP and classification algorithms would enhance the ability to filter out suspicious e-mails and would prevent malware infection. If, however, the malware is still able to infect the victim's machine and starts its execution, ML models can be deployed to perform runtime process analysis to detect malicious processes. As the malware scanned the

victim's machine to find web application credentials, ML-based AD systems or classification algorithms could efficiently distinguish an anomalous process from regular ones. In our view, unsupervised ML approaches would be more suitable for detection as obtaining labeled datasets for this type of scenario is rarely possible.

Step 2: Connecting to Target using stolen credentials

After stealing the web application credentials, the attackers were able to connect to Target's web services dedicated to vendors and run tests to expose vulnerabilities to exploit for the subsequent attack steps.

Non-ML Defenses: Non-ML techniques based on multi-factor authentication logins can be extremely effective in preventing this type of attack. Recall that the Target attack happened in 2013, and in recent years we have seen a clear transition to multi-factor authentication mechanisms.

ML Defenses: Once the attackers are connected to the server, ML techniques that leverage action sequence labeling can be deployed to detect users' anomalous behaviors. For instance, a possible solution is to employ unsupervised ML models that learn the typical users' behavior and recognize anomalous behavioral patterns. Classification and anomaly or intrusion detection algorithms can effectively perform such tasks. After detection, one could consider techniques (e.g., account blacklisting) to temporarily block the compromised account and prevent it from performing any additional actions.

Step 3: Exploiting a web application vulnerability

The attackers leveraged the stolen credentials to connect to the vendor services and discover existing vulnerabilities. In particular, the attackers discovered a vulnerability that allowed them to upload documents to Target's server without any check on the document type. Thus, the attackers uploaded a PHP script that generated a web-based backdoor on the server, and exploited the backdoor to upload additional files and execute arbitrary OS commands.

Non-ML Defenses: Techniques, such as rule-based filtering, can be effective for avoiding this type of vulnerability, provided that the rules are accurate.

ML Defenses: ML classification techniques (e.g., clustering and SVM) for filtering would be also suitable to identify the files' type. As, however, attackers could be able to evade the filtering, additional protection would be the deployment of ML techniques for runtime process analysis to detect anomalous processes. The reason is that the attacker process connected to the backdoor continuously executed OS commands with a behavior that likely deviated from the typical processes running on Target's servers. Unsupervised classification and AD algorithms can be viable options in this case.

Step 4: Searching relevant targets for propagation

By executing arbitrary commands, the attackers were able to perform reconnaissance campaigns to collect information on Target's internal infrastructure and identify potential targets to continue the attack. In particular, the attackers queried the active directory, which contained data on all entities on the internal network (e.g., users, computers, services, and SQL servers). Note that any user could query the active directory, and no special restrictions were applied to the query type. Thus, the attacker could retrieve the names and IP addresses of relevant SQL servers and point of sales (POS) machines.

Non-ML Defenses: One of the main vulnerability points lies in the fact that the active directory allowed each user to perform any type of query. Thus, an effective solution would have been to deploy access control for query execution, to limit the types and numbers of queries users can do on the directory. The reason is that typical users only perform a few queries against active directories, and only specific user issue queries (e.g., administrators). Therefore, one could issue fine-grained permissions to users to limit access to the directory.

ML Defenses: An alternative approach would be to use ML-based models of the user's normal behavior, possibly grouping users by roles or groups, and then carry out AD against the normal behavior of these roles/groups.

Step 5: Stealing access token from domain administrators (DAs)

After identifying relevant targets, the attackers needed a DA privilege to access them. Thus, the attackers employed a well-known technique called *Pass-the-Hash* to log in as a valid user. The attackers could steal a DA's NT hash (a token used in place of the actual password) from the Web server memory and use it to log in successfully.

Non-ML Defenses: Once again, the attackers could steal and exploit valid users' credentials. For this type of vulnerability, multi-factor authentication techniques, such as *proof of possession* for the token, can effectively prevent unauthenticated logins. Moreover, one can consider decreasing the expiration time associated with password tokens to reduce the time window in which each token can be used.

ML Defenses: Although detecting stolen credentials can be challenging, one can resort to strategies to analyze the users' behavior and detect users performing unexpected actions. As discussed in previous steps, one can consider ML algorithms for classification and AD.

Step 6: Creating new DA account using the stolen token

To ensure long-term control within Target's network, the attackers could not rely on the stolen admin credentials as these could expire or change. Therefore, the attackers leveraged the stolen credentials to create a new DA account. This step was highly critical as the attackers were able to obtain complete control over a domain admin account, which entails the maximum privilege permissions.

Non-ML Defenses: Given the permissions associated with admin accounts, one should consider a multi-step authorization mechanism every time a new admin account is created. Non-ML techniques can successfully implement this type of mechanism.
ML Defenses: Once an administrator account is created, one can consider ML techniques to learn typical admins' behavioral patterns and recognize anomalous ones. Once again, possible ML-based approaches include classification and anomaly detection algorithms.

Step 7: Propagating to relevant computers with new admin credentials

At this point, the attackers had information regarding the servers with relative information they want to access (from step 4) and the credentials to access these servers (from step 6). In order to propagate to the relevant computers with admin credentials, the attackers had to bypass the firewall. This was achieved by propagating through a series of servers and running a specialized tool to identify which computers were network accessible from the current computer.

Non-ML Defenses: Non-ML techniques can be used to impose fine-grained access control policies.
ML Defenses: Similar to step 6, ML techniques for AD can be used and combined with multi-factor authentication techniques.

Step 8: Stealing 70M of Records with Personal Identifiable Information (no credit cards)

With access to computers with databases of user information, the attackers tried to access Personal Identifiable Information (PII) of 70M users. The attackers tried to use bulk SQL copy tools, but that step failed due to Target being compliant with Payment Card Industry Data Security Standard (PCI). Accordingly, the attackers had to change their strategy and move to installing malware on POS.

> **Non-ML Defenses:** PCI compliance was in place and prevented the attack.

Step 9: Installing malware on POS and steal 40M credit cards

This is the first step in the attackers' alternative plan, which they switched to after failing to extract the credit card information from the SQL database. The alternative plan was to use the same propagation techniques in step 7 to install a malware on all POS machines. The malware scanned the memory of each POS machine to find credit card information, and then saved the collected information into a file on the machine's permanent storage.

> **Non-ML Defenses:** Techniques such as hardware memory encryption can be used so that memory contents for one process are not accessible by any other process nor by the operating system (e.g., Intel SGX/TDX). Such techniques are however expensive because they require specialized hardware and also may introduce delays.
> **ML Defenses:** ML techniques can be used to identify executable files that might be harmful and prevent their execution. Also, AD can be executed to detect unusual process behavior such as full-memory scanning. These techniques would not require specialized hardware. However, since the system under attack was a POS, one may not have been able to use a large and resource/time-consuming ML model. Accordingly, for this attack, the models should have had a very small memory footprint and very low inference latency as a POS is a customer-facing system. An additional defense is to use ML-based NLP scanning programs to classify files containing sensitive data and delete or prevent access to these files. Again, very lightweight models should be used here to minimize resource requirements and user-observable latency.

Steps 10 and 11: Sending and storing stolen data within Target's internal network; then sending stolen data via file transfer protocol (FTP) to an attacker-controlled server

In this step, the attackers copied the locally stored files containing credit card information to a remote server using standard FTP. However, the data was still within Target's internal network at this point. Afterwards, the attacker exfiltrated the data via FTP to an external attacker-controlled server.

> **Non-ML Defenses:** Techniques such as access control policies can be deployed to prevent FTP data transfer from a POS to any internal or external server, or between any internal server and an external server.
> **ML Defenses:** Once the data transfer starts, one can use network sniffing to analyze the packets and ML/NLP classification techniques to detect sensitive information in the packet such as credit card information.

8.2   The SolarWinds Attack

The SolarWinds attack is a supply chain breach that involved SolarWind's Orion system, a network management system tool used by more than 30,000 public and private organizations. The cybersecurity company FireEye first detected the breach. The company identified the inclusion of malicious code in the Orion build cycle, which gives hackers access to its systems. FireEye labeled this backdoor as "Sunburst". Alternatively, the backdoor is referred to as "Solorigate". We will use the term "Solorigate" for consistency. Extensive analysis has been conducted on the SolarWinds attack, which has been shown to be one of the most sophisticated attacks. The breach was discovered a year after the hackers gained unauthorized access to the SolarWinds network. In this section, we analyze the SolarWinds attack and propose ML-based and non-ML-based solutions that could thwart such an attack in different stages.

The attackers initially accessed SolarWinds's update server. This server compiles Orion's source code and distributes it to SolarWinds' customers. Then, they installed the SUNSPOT malware on the server, which compromised the compilation pipeline of Orion's source code by adding malicious compartments. The malicious code, incorporated in Orion's core, gave remote access to any customer network systems on which Orion was deployed. Once the binary was compiled, it was distributed to the customers. All the instances of Orion's software in the customer networks communicated with attacker command and control servers.

Preparation Step 1: Getting access to SolarWinds's network

The attackers had installed the SUNSPOT malware remotely. The question is, how were they able to do that? SolarWinds reported that an easily guessable password for the company's critical update server was stored in a private GitHub repository for more than a year. The CEO mentioned that the intern assigned the access password "solarwinds123" to the company's update server and posted the password to a private GitHub repository which several people may have access to. Presumably, this could have allowed the attackers easy access to the company's update server.

> **Non-ML Defenses:** The adoption of a strong authentication mechanism is highly recommended in critical environments such as an update server. Such authentication techniques could be two-factor authentication and public key-based authentication.

Preparation Step 2: Escalating privilege

At this point, SUNSPOT malware had been deployed on SolarWinds' production server. First, the malware created a mutex to ensure that only one instance would run in the system. Otherwise, multiple parallel executions of the malware may have risked detection. It then generated a log file for any errors that the malware encountered and deployment information. The log file would report information to the attackers, such as where it failed or the location of Orion's source code. In addition, the malware granted itself debugging privileges, meaning it could practically access the memory locations of other system processes.

> **Non-ML Defenses:** In critical production environments, system processes and their permissions must be documented. In other words, it should be clear which processes are allowed to execute and with what permissions. In addition, privilege elevation should not be permitted, at least from processes that should not have this ability. Such constraints could have made this step impractical to perform.

Preparation Step 3: Monitoring the running process to inject the code

Since SUNSPOT now had debugging privileges, it could perform system calls to check the current status of other processes. The malware was actively looking (i.e., every 1 s) for "MsBuild.exe", which is part of Microsoft Visual Studio development tools and was invoked when Orion was ready to build. If there was an "MsBuild.exe" process, SUNSPOT would spawn a new thread to determine if the Orion software was being built and then would hijack the build operation to inject Solorigate. The monitoring frequency of 1 s would allow the malware to essentially never miss any Orion's build operation and then hijack the process before sending the source code to the compiler. Due to the use of mutex (from step 2), the malware prevented itself from executing multiple monitoring loops. Threads were extremely risky since any compilation issue would have triggered manual inspections from the engineers. Thus, the attackers had to be very careful and stealthy.

> **Non-ML Defenses:** Monitoring other processes requires the invocation of certain system calls. One could whitelist the processes that should be allowed to perform such system calls while blocking the ones that should not.

> **ML Defenses:** The invocation of such system calls is a typical malware behavior. A ML model could be trained to recognize such behavioral patterns and classify them as malicious.

Preparation Step 4: Extracting the command-line arguments from process memory

The malware extracted the command-line arguments for each running "MsBuild.exe" process from the virtual memory. The command line was then parsed to extract individual arguments, and SUNSPOT looks for the directory path of the Orion software Visual Studio solution.

Preparation Step 5: Injecting Solorigate source code

After finding the directory of the Orion solution, SUNSPOT replaced the source file content with a malicious variant to inject Solorigate while Orion was being built. The malicious source code was stored in encrypted blobs to evade static analysis. SUNSPOT decrypted the code and then replaced the source file content, which involved creating and deleting intermediate files. The attackers adopted several sophisticated techniques to ensure the backdoor had a persistent foothold in the system without being detected. For example, they used compilation directives to remove any warnings generated at the output. The malicious code was added to a frequently invoked class, so the backdoor was always alive.

> **Non-ML Defenses:** Enforcing trusted execution environments by forcing the memory associated with the compiler to be read-only and (or) creating enclaves which are secure partitions within the main system memory would be effective defenses. The applications running in enclaves are encrypted, making them impenetrable to outsiders. **ML Defenses:** SUNSPOT shows several behaviors common in malware infections, including memory accesses to other process regions, creation of intermediate files, and deletion of system files. Hence, ML-based malware detection techniques can be helpful here.

Infection Step 1: Checking the victim's environment

Once the attack against the Orion's built was successful, the update server dispatched the compromised version to SolarWinds' customers. The customers are called the "victims" since those were the ultimate attack targets.

Solorigate carried out several checks to verify that there were no running processes related to security-related software (e.g., Windbg, Autoruns, and Wireshark) or that no analysis tools were present. Those checks included checking process names and file timestamps. If Solorigate detected such processes, it attempted to disable them by manipulating their configurations in the Windows Registry.

In addition, the malware checked the active directory domain in the victim's environment. The active directory naming convention may have revealed information about the network environment, such as if its local network is a demilitarized zone (DMZ). In such cases, the malware did not take any action. Finally, it checked for Internet connectivity to ensure that a command and control channel could be established.

> **Non-ML Defenses:** Since Sunburst did not re-check once it disabled the security-related service, checking unintentionally disabled services and restarting them could have mitigated this attack.
>
> **ML Defenses:** One could use ML models for recognizing suspicious activities such as disabling security services through Windows Registry. An ML model could be trained using features such as system calls to check security-related software and drivers and system calls to disable processes via Windows Registry.

Infection Step 2: Setting command and control

Solorigate used an intermediary command and control (C2) coordinator as a DNS server to retrieve its final C2 server. Solorigate used a domain generation algorithm to construct subdomains used to reach and exchange information with the C2 coordinator. Regular DNS queries and responses between the victim and the C2 coordinator were used to exchange encrypted information. The information and instructions for the malware were encoded in the IP ranges in the DNS responses. The C2 coordinator then communicated the victim-specific C2 server to Solorigate. Solorigate and the specific C2 server then communicated through encoded HTTP requests and responses.

> **Non-ML Defenses:** A whitelisting domain approach could have been sufficient to prevent the software from contacting unknown domains. This requires, however, knowledge of the trusted domains that Orion should be able to communicate with.
>
> **ML Defenses:** Since Solorigate performed queries to several subdomains of an unknown domain, AD techniques could detect such malicious behavior.

8.3   The WannaCry Ransomware

The WannaCry ransomware attack was a worldwide attack that happened in 2017. This attack targeted the Microsoft OS and encrypted the user data on affected machines. WannaCry exploited a Windows vulnerability in the server message block (SMB protocol). According to the estimate, it affected over 200,000 computers across 150 countries, with total damages ranging from hundreds of millions to billions of dollars.

In what follows, we introduce the different steps according to which the WannaCry attack was organized and discuss both non-ML and ML defenses that could have been used to avoid/mitigate the attack.

Step 1: Finding and infecting machines

To infect machines, WannaCry first found machines from both the internal network and the external network. It performed network probing within the internal network to find the IPs of machines to which it could connect. To spread over the external network, it randomly generated public IP addresses and performed a trial and error to find reachable machines. Once a list of reachable machines was found, WannaCry built connections to them and performed kernel exploitation. It used a vulnerability in the Windows OS called EternalBlue. EternalBlue was present in the Windows implementation of the SMB network protocol, and it had been originally developed by the US NSA and later leaked by a hacking accident. EternalBlue, in general, is an exploitation that uses a buffer overflow bug and then performs arbitrary heap write, therefore gaining control over the infected machine.

Non-ML Defenses: Traditional network monitoring techniques could be effective in detecting the large amount of network traffic generated by WannaCry. Intuitively, the host making such network requests should be forbidden from accessing the network before it is confirmed to be safe. Second, proper kernel sandboxing techniques could be adopted to minimize the privileged access of different system components. Ideally, when a subsystem of the kernel is compromised, other subsystems should not be affected.

ML Defenses: ML techniques can be used to improve network monitoring. Traditionally, a network alert is generated when certain traffic patterns are observed. However, these patterns are usually written manually and can be easily bypassed by skilled attackers. ML models can be trained on datasets consisting of network traffic traces and labels (whether there is an attack or not) to detect malicious events. However, effective and efficient ML is challenging because of unbalanced training datasets, as most real-world network traces are benign and do not contain any malicious traffic.

Step 2: Establishing persistence

Once WannaCry gained control over the infected machines, it tried to persist and hide itself so that it could not be easily detected and removed. To do so, it changed several system settings. First, it created a process called "Microsoft Security Center Service" to disguise itself as a legitimate system service. Next, it modified the Windows Registry table to make sure that it was launched automatically after every boot. The AutoRun feature provided by Windows was also used to establish memory persistence. Additionally, WannaCry granted itself full-disk access on the machine, which was later used to encrypt the user data. It then deleted all the existing backups, which prevented users from recovering the data, and disabled system security features, such as safe mode features.

> **Non-ML Defenses:** WannaCry persistence is achieved by operations with high-security impact. Thus, traditional auditing techniques to detect such suspicious operations and alert the system administrators could detect the ransomware. For instance, deleting system backups rarely happens in practice.
>
> **ML Defenses:** ML techniques can be used to analyze the operation sequences executed in the system. However, how to represent the system operation sequences to support effective learning remains challenging. Since the auditing log is often long and complex, ML can hardly learn anything if the input is not carefully built. The current trend is to leverage NLP techniques and graph models to learn the system logs both in texts and graphs.

Step 3: Loading the configurations

WannaCry loaded the "XIA resource" which corresponds to the password-protected ZIP file after going through the persistence phase. It dropped the file into the working directory loading the configuration data into memory. There were three bitcoin addresses available to choose from, one of which was written back to the configuration data. The address was used as the payment address in the extortion message demanding the ransom. After selecting an address, WannaCry set the hidden attributes for the working directory and granted full access to the target system for all the files. Next, the hardcoded public RSA keys were imported from the offset of *tasksche.exe*. The *t.wnry* contained the default encrypted AES required to decrypt the DLL responsible for file encryption on the infected machine. The first eight bytes of the *t.wnry* were checked for the string WANACRY! and decrypted the AES key stored at the beginning of the *t.wnry* file using the imported RSA key.

> **Non-ML Defenses:** A static analysis can be carried out by disassembling the ZIP files for identification of the files containing the encrypted keys, configuration data, DLLs, and permissions.
>
> **ML Defenses:** Log analysis can be carried out on the infected host for malware activities, and TF-IDF can be applied to strings, patterns, and URL identification typical of ransomware. ML models can be used to detect abnormal log leveraging clustering algorithms.

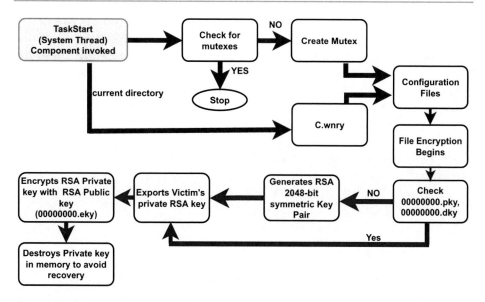

Fig. 8.2 Encryption component procedure

Step 4: Encrypting the files

After loading all the configuration data, the encryption component of WannaCry encrypted the files by following the procedure shown in Fig. 8.2. The procedure was invoked by the *TaskStart* thread and checked for one of the three mutexes. If a mutex were present, the task stops; otherwise, the encryption began with reading the content of *C.wnry* file containing Tor addresses from current directory and creating the "*MsWin-ZonesCacheCounterMutexA*" mutex . After that, WannaCry created three configuration files containing the information of TOR/C2 addresses, the public RSA key, and the encrypted private RSA key.

To encrypt the files, WannaCry checked the existence of the public RSA key (00000000.pky) and decryption RSA key(00000000.dky). The decryption RSA key was received after the payment verification. If the two keys did not exist, a RSA 2048-bit symmetric key pair was generated after which WannaCry exported the victim's private key and encrypted the private key with a hardcoded RSA public key (00000000.eky). Once the key was stored, WannaCry cleared off the memory to avoid the recovery of the private key. WannaCry kept track of the logical drives attached to the system and navigated the directories to encrypt the files of interest by generating a 16-byte symmetric AES key. The generated AES key was encrypted with the public RSA key and stored in the file starting with WANACRY! string and extension *.wncry*.

Non-ML Defenses: The use of decoy files can help in tracking accesses to the files by malicious processes. Decoy files are used to deceive DLL search order hijacking by spoofing the shared file libraries. File analysis is also important for analyzing and listing the shared file libraries and information regarding the permissions on the files. It requires disassembling the files and the associated DLL libraries. This further helps in whitelisting the processes that should be allowed to access the files and restricting unauthorized access.

ML Defenses: ML-based analysis can be executed based on the sequence and frequency of I/O operations, API calls, and opcode for detecting malicious behavior. Packers and encryptors can be detected by applying NLP-based techniques integrated with static and dynamic analyses.

Step 5: Preventing recovery

After completing the encryption, the goal of WannaCry was to avoid the recovery of data from the memory. Some of the strategies it adopted include

- Deleting all the shadow volumes without alerting the user.
- Ensuring the machine reboots even if any error occurs.
- Disabling the Window recovery feature, thereby, preventing users to revert their systems.
- Ensuring that the victim was unable to use any backup files created by the Windows server.

Non-ML Defenses: To monitor the memory and restore memory back to its normal state, memory snapshots can be taken at regular intervals. The snapshots can also be helpful for carrying out memory forensics which gathers the memory logs and memory dumps to identify the abnormalities created by the malware.

ML Defenses: Memory forensics can be carried out to fetch volatile memory dumps and patterns and, thereby, can be used to create meta-features to train the ML model for analysis. The snapshots taken can be used to fetch API function calls, registry activities, and imported libraries to apply ML-based techniques for analysis.

Step 6: Propagating

WannaCry used the EternalBlue exploit and the DoublePulsar backdoor to leverage the MS17-010 SMB vulnerability. It obtained the IP addresses of local network interfaces and the worm component tried to connect to all possible IP addresses in any available local network on port 445.

Non-ML Defenses: The basic solution could be to block the IP address of the infected machine and isolate it from the network to prevent further scanning for propagation. Apart from isolation, limiting the number of connections any host can make in the local as well as a public network can also be useful. Another defense is to list the unused incoming open ports, disable them, and allow the sharing of files from secured ports like ports 8443, 4500, etc. SMB vulnerability was present in the old version of Windows. Installing a newer version could make the attack more difficult as the attackers would need to find other exploitable vulnerabilities.

ML Defenses: ML-based network behavioral analysis can be carried out to identify port scanning, suspicious connections, unauthorized file transfers, infected IPs, and so on. This also includes abnormal traffic detection.

Step 7: Communicating

WannaCry unpacked and dropped the files from the s.wnry file, which contained the Tor executable, into the installation directory. If the contents of the s.wnry file were corrupted, WannaCry would download the Tor executable from a hardcoded URL. WannaCry parsed the contents of the c.wnry file, which specified the configuration data. During its communication with Tor addresses, WannaCry established a secure HTTPS channel and used Tor ports for network traffic and directory information.

Non-ML Defenses: Application control mechanisms can mitigate illegitimate modifications to the system. A few rules can be implemented on the files like adding read-only files to user machines to prevent changes to the executables. URL filtering can be employed to identify hardcoded suspicious URLs.

ML Defenses: ML-based behavioral monitoring and network traffic analysis can aid to identify flow patterns, suspicious URLs, IPs exploited, ports utilized, and so on.

Challenges in the Use of ML for Security

9

The use of ML techniques for security has several challenges, many of which are common with other applications. Main challenges include training data scarcity and quality, proper model configurations, ML ethics and security. In what follows, we discuss these challenges and related proposed approaches, if any, and emphasize whenever appropriate a security perspective.

9.1   Data Availability and Quality

In previous chapters, we have shown that ML techniques are highly effective for several security functions, such as malware classification and intrusion detection. Training ML models, however, requires vast amounts of labeled data. Obtaining abundant labels is the main obstacle in applying ML techniques for security, due to the high cost of data labeling and the evolving nature of attacks. In what follows, we discuss four issues related to data acquisition commonly encountered across security functions.

- *Scarcity of labeled data,* which refers to when the amount of labeled data does not suffice to achieve satisfactory training results. Attack data is often very time-consuming and costly to collect and label. Investigating attack samples requires substantial domain expertise and effort. Consequently, only smalls sets of data can be labeled manually by security analysts.
- *Imbalanced classes,* which refers to the imbalanced coverage of all class labels. Take network intrusion detection for example; a labeled dataset might have little or no samples for new attack types, while having a lot of samples for old attacks. Imbalanced data is a

© The Author(s), under exclusive license to Springer Nature Switzerland AG 2023   131
E. Bertino et al., *Machine Learning Techniques for Cybersecurity*, Synthesis Lectures
on Information Security, Privacy, and Trust,
https://doi.org/10.1007/978-3-031-28259-1_9

notorious issue which causes model performance degradation, since a classifier cannot accurately detect all types of classes.

- *Coarse-grained labels,* which refers to when the provided labels lack specificity. For example, for malware classification, evolving malware of the same family may be assigned to a generic label. However, they may use different tactics and techniques in initial infection and lateral movement, hence based on different kill chains. For timely and targeted post-infection actions, each malware should be assigned a specific label.
- *Noisy labels,* which refers to when the analyst assigns incorrect labels by mistake. In malware detection, different attack families may be classified under the same family. We differentiate the noisy labels from data poisoning attacks introduced in Sect. 9.4 where the attacker inserts carefully crafted malicious samples into the training dataset.

A promising approach to address label scarcity is the use of TL techniques by which knowledge, in form of a pre-trained model or training data, can be transferred from one domain, referred to as source domain, to another domain referred to as target domain that has scarce training data. Conventional TL-based approaches usually leverage a pre-trained model and fine-tune the trainable parameters using limited training samples in the target domain. However, these pre-trained models typically learn inferences from a huge dataset and consequently, the models contain many redundant features or irrelevant latent spaces that have no benefits to the target inference tasks. In addition, they require manual efforts, to decide for example which layers are trainable. Adversarial domain adaptation, on the other hand, aims to learn the target task by leveraging some training samples from a source domain that is related to the target domain, but, at the same time, attempts to directly minimize the discrepancy in latent space distributions between the source and the target. We refer readers to the approach by Singla et al. [215], which uses adversarial domain adaptation to address the problem of scarcity of labeled training data in network intrusion detection datasets (see also Sect. 2.7.3).

With respect to the data imbalance issue, common approaches include weighted loss functions, or oversampling and undersampling of the training data in the minority and majority classes, respectively. However, the effectiveness of these methods highly depends on the nature of the datasets and the learning task at hand. For image data, another category of techniques is to leverage unsupervised learning algorithms, such as the variational autoencoder [212], and produce synthetic images for data augmentation. Unfortunately, such techniques cannot be directly applied to security application data because they are non-image data.

To address fine-grained attack classification under coarse-grained and noisy labels, the approach by Liang et al. [121] combines different clustering algorithms with the given labels to mitigate the negative impact of low-quality labels. The input of the approach is a dataset with limited labeled data. The output is a clustering assignment for all the data samples. The data samples are expected to be either correctly clustered under the known labels or form new clusters representing new labels. These clustering results can then be used by human analysts to derive labels efficiently.

9.2   Selection of Models, Hyperparameters, and Configurations

Selecting the right ML model and its hyperparameters is an essential and challenging step to achieve the best model accuracy. Additionally, the model type and hyperparameters can significantly impact the model's training and inference times. In general, the higher the complexity of the model, the higher the number of the model's parameters and hence the longer its training time. Although the majority of model and parameter selection decisions are based on mere exploration, there are a few semi-automated techniques, such as grid search, random search, and Bayesian optimization, that can narrow down the options using the nature of the problem and the dataset.

9.2.1   Selecting the Right Model

Selecting the right model is usually achieved through domain knowledge followed by a systematic evaluation. Depending on the nature of the problem and the dataset, the type of the right model can vary. For example, if plenty of labeled data points are available for all classes, then a supervised model is adequate. However, for data that have no labels, or for which collecting labels is either costly or impractical, an unsupervised learning algorithm is more appropriate. One example of such a domain with very few labeled points is outlier detection, in which there are usually none or a few samples that represent the outlier class. In this case, a clustering algorithm is usually used to identify the boundaries for non-malicious applications, and any application that falls outside of these boundaries is flagged as an outlier.

On the other hand, if the system setup allows for a domain expert to provide feedback (i.e., human in the loop), then using a semi-supervised model is more adequate. In this case, unlike supervised models, only a few labeled data points are sufficient. One example of such a domain is malware detection systems. Here, a few *suspected* malicious applications can be manually evaluated by a domain expert, and then their label is sent back to the model for refinement.

9.2.2   Hyperparameter and Configuration Tuning

After selecting and evaluating the right model type, it is very critical to select the right hyperparameters for the model to achieve the desired accuracy. Hyperparameters control the complexity, and in many cases, also control the model's vulnerability to overfitting. Examples of hyperparameters are (1) number of layers/neurons in a deep learning model; (2) number of trees in a random forest; (3) number of clusters in k-means clustering. Even for the same dataset, different values of the hyperparameters can yield significantly different model accuracy.

The simplest technique to select the best hyperparameters is to evaluate all possible values and select the one that yields the best model accuracy. However, this is usually an impractical technique due to several reasons: (1) the abundance of the hyperparameters, making the combinatorial search space very large; (2) the continuous and unbounded nature of the hyperparameters (e.g., number of trees); (3) the dependency between the hyperparameters. Accordingly, the most common technique for hyperparameter tuning is the grid search by which one splits each hyperparameter into a finite and discrete set of values, organizing the search space into a grid. For example, suppose we are tuning two hyperparameters for a DL model, and we need to determine the number of layers and the number of neurons in each layer. First, we select a set of reasonable values for each hyperparameter such as layers = [2, 4, 6] and neurons = [10, 20, 30]. Second, we evaluate the model's accuracy with all possible combinations of the two parameters: [(2, 2, 2), (2, 2, 4)...(6, 6, 6)]. We finally select the combination that leads to the best accuracy.

Alternatively, random searching can be used to evaluate only a subset of the grid, which is usually a more appropriate technique when the hyperparameter tuning budget (in time or $ cost) is limited. Both grid and random searching are inherently parallel techniques, in which all points can be collected and evaluated in parallel with no dependencies. Additionally, Bayesian optimization (BO) techniques are also common in selecting the best hyperparameters. Similar to random searching, BO operates with a budget requirement on how many points to sample. However, BO is not inherently parallel and searches the space in a one-by-one manner, but it has the advantage of finding near-optimal parameter values in very few runs due to its acquisition function. The acquisition function is executed after each data point collection, and it decides where in the space to sample the next point. By selecting an appropriate acquisition function, BO can provide better hyperparameters over random or grid searching under the same sampling budget.

9.3   Ethics

As discussed in the seminal paper by Cowls and Floridi [57], the fact that AI will have a major impact on society is no longer in question. However for AI technology to deliver its full potential benefits is critical that it is not underused—typically for the wrong reasons, thus limiting its applications, or overused and misused, thus creating risks. Concerning the risks, it is important to notice that everything from email phishing campaigns to cyber wars may be accelerated or intensified by malicious use of AI [221]. In the area of cybersecurity, a notable example of ML misuse is represented by gendered differences in face recognition [5], which can result in misclassification when using images for biometric authentication. Other issues are related to unequal privacy-preserving protections for specific racial groups [14].

As a result, AI technology and especially ML techniques have been intensely scrutinized with respect to ethics [178]. Many initiatives to address the ethical use of AI have been

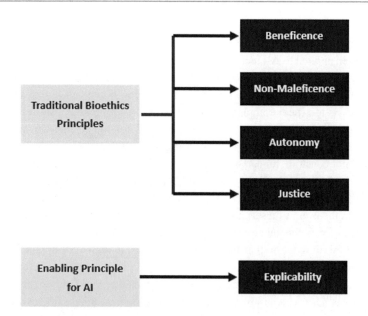

Fig. 9.1 AI ethical principles (redrawn from Fig. 4.1 in [57])

undertaken by many different organizations, including governments, civil liberties and no-profit organizations, professional organizations, and industry.

In general, there is an agreement on the foundational principles of AI ethics (see Fig. 9.1) [57]. Four of them, namely beneficence, non-maleficence, autonomy, and justice, are based on traditional principles from bioethics, whereas the other one, explicability, is a novel principle specific to AI. Beneficence requires that AI-based processes, such as decision-making processes, need to explicitly benefit the end-users. End-users must be the beneficiaries of AI-based processes, not organizations or other regulatory bodies. In other words, just like in medical care, a vaccine or other public health rule needs to benefit first and foremost the individuals that are asked to accept it, not the issuing organizations [52]. Conversely, any implementation/deployment of AI technology, including ML for cybersecurity, should not generate adverse side-effects that counter the beneficial aspects. The principle of non-maleficence states that if, for example, end-users are asked to share with a ML-driven process personal data for controlling access to certain resources, the data will not be used in further deployments of ML-based tools that may lead to surveillance of the same individuals. The principle of autonomy in the context of bioethics refers to the rights of individuals to make their own health decisions, such as about which medical treatments to receive, even though they often surrender such decisions to experts. With the deployment of AI technology, individuals (will) surrender some of their decisions to AI algorithms and models. As pointed out by Cowls and Floridi [57], based on an analysis of several documents, "not only should the autonomy of humans be promoted, but also the autonomy of machines should be restricted and made intrinsically reversible, should human autonomy need to be re-established." The justice principle, even though intuitively clear, has some variations when applied to AI, including the promotion of social justice and the elimination of all types of discrimination, shared benefit and shared prosperity, and solidarity. Finally, the principle of explicability, by which Cowls and Floridi synthesize explainability and accountability, is a critical complement to the other four principles. As well exemplified by Cowls and Floridi [57] "for AI to be beneficent and non-maleficent, we must be able to understand the good or harm it is actually doing to society, and in which ways; for AI to promote and not constrain human autonomy, our 'decision about who should decide' must be informed by knowledge of how AI would act instead of us; and for AI to be just, we must ensure that the technology—including its human developers and deployers—are held accountable in the event of a serious, negative outcome, which would require in turn some understanding of why this outcome arose."

It is important to mention that those principles, even though very important from conceptual and foundational points of view, are still very difficult to apply in actual AI systems and tools. Therefore, industry and research in technical fields have focused on "technical AI ethics principles", for which implementation and deployment in AI systems are promising and which are being widely investigated from different perspectives by researchers, companies, and governmental organizations. These principles include explainability, fairness, robustness, transparency, and privacy. As they are very relevant for the application of ML to security, we discuss them in what follows.

9.3.1 Explainability

Explainability, also referred to as interpretability, can be loosely defined as how well humans can understand the decisions of a ML model. It can be characterized by the following different perspectives [109]. From the global explainability perspective, the elements that need to be explained cover all the steps of the ML processing chain, including (i) The features used in the model; it is also critical that these features be interpretable and conveyed in a way that humans can easily understand (we refer the reader to [258] for details on the notion of interpretable features). (ii) The types of data expected to be used in the model, including the boundaries of input space (e.g., only valid for students aged from 10 to 15, or for images taken during the night). (iii) In the case of the classification tasks, how the decision is taken using the output values (e.g., in the case of thresholding, how the threshold is chosen) [109].

The second perspective focuses on explaining a single prediction made by the ML model using specific input data. While such an explanation can be based on the understanding of the global model, specific approaches can be devised to explain the specific decision. Approaches include highlighting the features that had the major weight in the decisions based on the notion of quantitative input influence [61], or providing counterfactual explanations [236]. Counterfactual explanations refer to a statement of the form: "If X had not occurred, Y would not have occurred". They suggest what should be different in the input instance to change the outcome of a ML model [94] and are considered quite effective. As an example, consider a bank loan request that is rejected. A counterfactual explanation would be a statement like: "if the income had been 1500$ higher than the current one, and if the customer had not had unpaid debts with other banks, the loan would have been granted."

ML explainability for security is critical for several security tasks. For example, when using a ML model for ABAC decisions, it is important to determine why a given subject was (or was not) given access to a certain resource, or when using a ML for a network anomaly detection to determine which features are relevant and which ones are not for detecting malicious packets in order to reduce the overhead associated with anomaly detection. However to date, there is limited research addressing specific issues related to explainability for security. One notable example is related to the use of GNN for analyzing attack graphs in the context of forensics (see Sect. 7.3).

We refer the reader to [82] for a detailed tutorial on explainability and to [170] for additional resources, including survey papers and tools.

9.3.2 Fairness

In the context of decision-making, fairness can be defined as "The absence of any prejudice or favoritism toward an individual or a group based on their inherent or acquired characteristics." [156]. These prejudices/favoritism are also referred to as biases. Concerning the notion of fairness for ML, it is important to notice that this notion has many different perspectives; we refer the reader to [189] for a detailed discussion of these perspectives.

Because ML is today used in many human/socially sensitive decisions, it is critical to ensure that these decisions do not reflect discriminatory behavior toward certain groups or populations. Over the years, cases have emerged showing that biases in data/algorithms resulted in discrimination; we refer the reader to [156] for the analysis of one such case. The problem of ML fairness has been widely investigated with respect to (i) the different types of bias [189]; (ii) which types of bias is introduced in the data, algorithm, and user interaction ML feedback loop [156]; and (iii) "fairness tools" [205]. It is important to notice that because of its many dimensions, the ML fairness problem is quite complex and thus mainly specialized solutions, including tools, have been proposed—some of which are specialized for different ML techniques. Just to give an idea of the complexity, Fig. 9.2 shows the various types of bias in ML. Pre-existing biases "are associated with an individual or institutions. When human preferences or societal stereotypes influence the data and/or the model, that is said to be the effect of pre-existing bias." [189]. These biases are then reflected in data collected over time. Technical biases are the consequences of the limitations of computer and data technology and how the technology might introduce biases. For example, the selection of features, models, and training procedures can introduce bias that is not affiliated with the party performing the training but with the inadequacy of those procedures

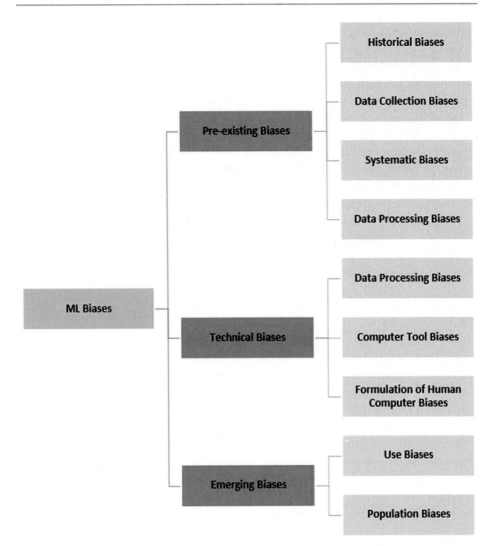

Fig. 9.2 A taxonomy of type of bias in ML (based on Fig. 1 in [189])

to characterize [189]. Finally, emerging biases are biases that result from the deployment
of a model. Emerging bias can emerge in two different forms, via population bias and use
bias. Population bias stems from the insufficiency of a model to represent its population or
society after model deployment, whereas use biases—which are an interesting category of
biases—refer to biases that humans acquire after using a model [189].

ML fairness is relevant for security, especially in the context of access control, to make
sure that users are given fair access to controlled resources. It is also relevant for privacy—for
example to ensure that different population groups have the same level of privacy protection.

However, it is clear that, given the complexity of the fairness problem, specific fairness strategies must be devised and analyzed for the specific security task at hand.

9.3.3 Robustness

Robustness refers to the ability of a trained model to do well when there are changes in the environment where the model is deployed. There are various reasons for these changes, including malicious attacks, unmodeled phenomena, undetected biases, and significant changes in data [246]. As discussed by Woods [246], achieving model robustness is challenging in that it requires addressing several issues, including

- *Data quality*, which at a high level refers to the data being "fit" for the task at hand. Data quality in turn has many different dimensions, including correctness, consistency, accuracy, timeliness, and no presence of biases. Data quality has been widely investigated over the past 20 years and approaches and methodologies have been proposed, some of which are based on ML. We refer the reader to the book by Batini and Scannapieco [26] for a comprehensive coverage of the data quality area. We notice however that data quality is very much application-dependent and thus good domain knowledge is critical to ensure the quality of the data used for training ML models.
- *Model decay*, also called dataset shift, which refers to when new data distributions can arise that differ from historical distributions used to train models. Addressing model decay requires re-training the model with new data, which can be a problem if dataset shifts are frequent, as one may not have a sufficient amount of new data to re-train a model. An approach to address this issue is based on domain adaptation techniques (see Chap. 2) that allow one to adapt the older training data in order to combine them with the new data.
- *Feature stability*, which refers to variations in the relevance of the features used in the trained model. Stable features typically perform well for a large variety of input data. Approaches have been proposed for evaluating feature stability and we refer the reader to [181] for a comprehensive survey of these approaches.
- *Precision versus Recall*, which refers to identifying a good trade-off between these two metrics. More specifically, "Precision is a measure of exactness in model operation. Recall is a measure of completeness or quantity. 'High precision' means that a model returns substantially more relevant results than irrelevant ones. 'High recall' means that a model returns most of the relevant results that are available. There is often a trade-off between precision and recall during model training, and practitioners must determine what balance is best for any given case. Getting the balance wrong may impact the robustness of the model" [246].

Robustness is critical in the use of ML for security for two reasons. The first is that attackers may manipulate input data to bypass models used for defense, such as intrusion detection. The second is that many novel environments that need to be secured, such as edge computing systems, cellular networks, and IoT systems, are very dynamic. In both cases, robustness and also approaches to quickly and continuously re-train the ML models are essential.

9.3.4 Transparency

Transparency can be defined as the ability of subjects to effectively gain access to all information related to data and processes used in decisions that affect them [29]. Such a definition covers many different types of subjects: data participant, a subject whose (personal) data have been collected and possibly used; data victim, a subject affected by the use of certain datasets or processes—we note that a subject can be victim[1] even though their data has not been collected/used; data user, a subject using data/processes for decision, research, etc. In the context of ML transparency, four different dimensions are relevant:

- Training data transparency—it refers to information concerning the training dataset collection, including the context (time, location, and so on) of data collection; and the data collection agent (applications, sensors, and human users).
- Model transparency—it refers to all information related to the training of the model, including hyperparameters used, thresholds, etc. We note that the notion of model transparency is closely related to the notion of global explainability.
- Model use and provisioning—it refers to information about processes/decisions for which the model has been used and to parties that have provisioned the model.
- Law and policy transparency—it refers to making all laws, regulations, and organizational policies associated with the use of ML models available to subjects.

It is important to mention that full transparency may conflict with business confidentiality and personal privacy requirements. Therefore, proper trade-offs must be devised, based perhaps on the use of aggregate/anonymized information.

In the case of security, transparency is important to allow users, such as security staff, to understand for example the reasons why a ML model has not detected a malware or a malicious network flow, and more in general to carry out forensics of ML models.

---

[1] We use the term victim to emphasize that, in most cases, subjects are adversely affected. However, subjects can also be positively affected and may actually be interested in getting information about the used data/processes.

9.3.5   Privacy

Privacy is a fundamental human right and has been widely investigated in relation to data, and it is today a research field more active than ever. As ML relies on huge datasets for training, ML privacy is critical. A specific threat introduced by ML is model inversion, which refers to the inference of data item values present in the training dataset of a model based on queries to the model. Many variations of model inversion exist, and we refer the reader to [154] for some recent work. Approaches have been proposed to address this problem, including the use of differentially-private training datasets, approaches that add perturbation during the training process, and training processes that work on encrypted data.

In security, it is critical to use privacy-preserving ML models when dealing with security tasks related to humans, such as access control and authentication, especially biometrics-based authentication. One should however carefully assess which privacy-preserving approach may be suited for the specific task at hand.

9.4   Security of ML

The widespread use of ML has pushed to the forefront the problem of securing ML models and algorithms. This problem is especially critical when ML is used for cybersecurity as one has to make sure that attackers are not able to evade ML-based defenses.

Because of its criticality, the problem of ML security has been widely investigated over the past 15 years. Several categories of attacks and related defense techniques have been identified. Attacks can be classified into two major categories: (a) data poisoning attacks that aim at manipulating the training data; (b) inference attacks that aim at manipulating the model inferences.

Data poisoning attacks include

- *Availability attacks*, also known as denial-of-service attacks, where the attack goal is to maximize the overall loss of the model and cause a degradation in model performance as well as misclassifications. Examples of these attacks are gradient-based and GAN-based attacks.
- *Integrity attacks*, where the attacker inserts carefully crafted malicious samples into the training dataset without affecting the model's classification of clean samples. Examples of these attacks are clean-label poisoning and backdoor attacks. Backdoor attacks—the most interesting attacks—only misclassify inputs containing specific (explicit or even implicit) triggers. A trigger example for a DL model used for face-based authentication would be the presence of glasses with a certain shape. Whenever the model analyzes a facial image with glasses with such a shape, the model would classify the image as the one of a given system administrator, thus allowing attackers to impersonate the administrator.

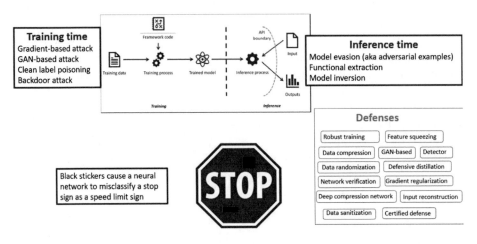

Fig. 9.3 ML attacks and defenses

Inference attacks include

- *Evasion attacks*, by which attackers make the ML model misclassify input instances. Let x be a clean sample, M a ML model, and y be the class predicted by M for x. The attacker adds an imperceptible perturbation δ to x to create an adversarial example $x' = x + \delta$, which is sufficient to mislead M to produce a wrong output y' different from y (untargeted attack). A variation of this attack is the targeted attack; the goal of the attacker is to mislead M to predict an attacker-targeted class y_{target}. A well-known example is the one by which an attacker adds a black sticker to a stop sign to make the model misclassify the image of the stop sign as the image of a speed limit. Those attacks have been widely studied and there are several variations, such as the ones based on the knowledge that the attacker has about the model (e.g., black-box, white-box, and gray-box).
- *Model inversion*, by which the attackers learn information about the data used for training the model. This type of attack is particularly relevant to privacy, when the training data are privacy-sensitive.
- *Functional extraction*, which allows an attacker to learn an approximate model of a model that is kept confidential.

To protect against those various attacks, many different defense techniques have been proposed (see Fig. 9.3 for a list of these techniques). It is important to notice that robust training (see Sect. 9.3.3) is one of the possible defenses; however, it is only relevant to model evasion attacks. For other attacks, different defenses need to be adopted. We refer the reader to [103] for a comprehensive survey of both attacks and defenses.

An observation about ML-based security is that, while focusing on specific attacks and defenses is important, we also need to look at a broader picture. Consider the case of adversarial example attacks [30]. Consider the well-known example of an autonomous vehicle running a DNN to recognize stop signs and an attacker that has partially covered a stop sign so that the network is not able to recognize it. An objection to such an example is that autonomous vehicles will most likely have maps on-board with stop signs marked in them; stop signs in the future may also emit sounds, and there will be vehicle-to-vehicle communications and information transmission among vehicles. In other words, vehicles will have multiple information sources that can be compared and correlated so that the correct and safe decision is taken. We should therefore look at the problem of ML security from a system point of view by which one can secure the system of interest by including different ML techniques and models, and using data from independent sources. Such an approach would not only enhance security but also the timeliness and coverage of decisions, predictions, etc. In other words, our ultimate goal should not be just the protection of ML itself; rather it should be to make accurate decisions, forecasts, and analyses and to achieve this goal we need to think in terms of systems security. A recent approach along those lines for network intrusion detection is by Lee et al. [142] (see also discussion in Sect. 7.2); under this approach, different ML-based detectors are deployed, specialized for different attack steps. For an attacker to evade this approach, the attacker would need to evade all the ML detectors, thus raising the bar for the attacker.

9.5   Research Directions

The problem of ethics in security has been primarily investigated with respect to security research. Today, security conferences have requirements such as responsible disclosure, by which authors of papers reporting certain vulnerabilities are responsible to inform the affected parties before the papers are published. Other aspects concern the privacy of data used in security experiments, when these experiments involve privacy-sensitive data. A recent workshop [75] has focused on ethics in computer security research with the goal of starting more systematic discussions and analyses. However, not much work has been done on ethics concerning specific aspects related to the deployments and use of security techniques. For example, approaches such as active defense and retaliation need to be investigated with respect to ethics, including ethics related to ML models/techniques that may be used as part of these approaches.

Concluding Remarks

10

Cybersecurity is today more critical than ever, given our society's increasing reliance on cyber infrastructures and data. ML, which during the past 15 years has grown in sophistication, performance, and application scope, represents an important technology for enhancing cybersecurity. As a result, ML has been widely applied to several security functions, especially anomaly detection, malware classification, and digital forensics. There is still limited research for other areas, such as attack management.

Concerning the specific ML techniques used in the various ML-based security functions, many different techniques are used, ranging from association rules to attention mechanisms, reinforcement learning, and graph neural networks. We can certainly expect that as novel techniques are being developed in the AI/ML area, these techniques will be applied to security functions.

However, despite the very encouraging state of the art in the application of ML to security functions, we still have a very long way to go. One important issue is related to the reproducibility and transparency of results using ML techniques. Most of the approaches that we have analyzed include experimental evaluations of the proposed approaches. However, results often depend on the datasets used in the experiments and on other aspects, such as model parameters. Therefore, evaluation results may not always be conclusive, and one should be very careful in actual applications. Also, designing and training a ML model may require a lot of manual effort to fine-tune the model and clean the datasets used for training; those efforts are not often measured and reported.

Finally, an important area of application of ML is related to security configuration and management. This is a critical area as many attacks are due to incorrect configurations and poor security management. Today, there is an increasing trend toward using configurable software, that is, software that can be customized without additional programming [85].

© The Author(s), under exclusive license to Springer Nature Switzerland AG 2023   145
E. Bertino et al., *Machine Learning Techniques for Cybersecurity*, Synthesis Lectures on Information Security, Privacy, and Trust,
https://doi.org/10.1007/978-3-031-28259-1_10

Configurable software typically exposes a set of configuration options which can then be combined in different ways, resulting in different instances of the software. However, configurable software often does not control the correctness of such combinations (see, for example, [95]). Other examples of misconfiguration are related to firewall rules and access control policies. Concerning security management, the main issue is that it is expensive in terms of human costs. A recent interesting example is represented by the zero-trust paradigm, which would require specifying a large number of policies, deciding where to place the policy enforcement points, and managing policy changes to reflect changes in application and network traffic patterns [133]. Deploying and managing zero-trust architectures would be extremely expensive in terms of human efforts. It is, thus, clear that automated frameworks for security configuration and management are critical. We believe that ML techniques will play an important role in the design of those automated frameworks.

Appendix: Publicly Available Datasets

In the appendix we provide a short description of the publicly available datasets that have been used for experimental analyses in some of the approaches described in this monograph.

© The Editor(s) (if applicable) and The Author(s), under exclusive license to Springer
Nature Switzerland AG 2023
E. Bertino et al., *Machine Learning Techniques for Cybersecurity*, Synthesis Lectures
on Information Security, Privacy, and Trust,
https://doi.org/10.1007/978-3-031-28259-1

Table A.1 Publicly available datasets

No	Dataset	Task	Description	Available on	Accessed on
1	Amazon Access Samples	Security Policy Learning	The data represent logs of access control decisions, both positive and negative. It is composed of two files. The first file contains users and their assigned access. The second file contains the access history for a given user –Number of Instances: 30000 –Number of Attributes: 20000	http://archive.ics.uci.edu/ml/datasets/Amazon+Access+Samples	Aug 11, 2022
2	Project Management	Security Policy Learning	The downloadable contains synthetic logs of ABAC policies. The source code with which the data are created is also provided. Please refer to the paper (https://link.springer.com/content/pdf/10.1007/978-3-662-43936-4_18.pdf) for more details on how the synthetic data were generated	https://www3.cs.stonybrook.edu/stoller/software/ABACMiningFromLogs.zip	Aug 29, 2022
3	Nobi et al. [172] Access Control Dataset	Security Policy Learning	The authors provide two real-world and eight synthetic datasets. Please refer to the paper for more details on how the synthetic datasets were created. The syntax of an authorization tuple is unique id of a user u l unique id of a resource r l metadata values of all the user metadata of user u l metadata values of all the resource metadata of resource r l access information of all the four operations	https://github.com/dlbac/DlbacAlpha/tree/main/dataset	Aug 29, 2022

(continued)

Table A.1 (continued)

No	Dataset	Task	Description	Available on	Accessed on
4	Amazon Employee Access Challenge in Kaggle	Security Policy Learning	The data is composed of employee historical access requests between 2010 and 2011. It is composed of a train set and a test set. Each row has the ACTION (ground truth), RESOURCE, and information about the employee's role at the time of approval	https://www.kaggle.com/c/amazon-employee-access-challenge/	Aug 29, 2022
5	DSOS	Security Policy Learning, IoT Anomaly Detection	The dataset contains normal and malicious IoT traces captured in DS2OS. The traces contain application layer data such as meaningful device identifiers, device locations, and requested operations along with any potential parameters	https://www.kaggle.com/datasets/francoisxa/ds2ostraffictraces?select=mainSimulationAccessTraces.csv	Aug 29, 2022
6	CICAndMal2017	Android Malware Analysis, Malware Detection	The dataset contain both malicious and benign Android apps with 43 malware families and 5,065 benign apps captured using real smartphones	https://www.unb.ca/cic/datasets/andmal2017.html	Sept 4, 2022
7	Fishler et al. [80]	Security Policy Learning	The dataset contains access control policy examples for a Collage expressed in XACML	http://www.margrave-tool.org/v1+v2/margrave/versions/01-01/examples/	Sept 26, 2022

References

1. Aghakhani, H., Gritti, F., Mecca, F., Lindorfer, M., Ortolani, S., Balzarotti, D., Vigna, G., Kruegel, C.: When malware is packin'heat; limits of machine learning classifiers based on static analysis features. *Proceedings of the 27th Network and Distributed Systems Security Symposium (NDSS'20)*, 1–20
2. Ahmed, M., Mahmood, A. N., Hu, J.: A survey of network anomaly detection techniques. *Journal of Network and Computer Applications*, 60, 19–31
3. Ahmadi, M., Farkhani, R., Williams, R., Lu, L.: Finding bugs using your own code: detecting functionally-similar yet inconsistent code. *Proceedings of the 30th USENIX Security Symposium (USENIX Security'21)*, 1037–1054
4. Aich, S., Sural, S., Majumdar, A.K.: STARBAC: Spatiotemporal role based access control. *Proceedings of the 2007 OTM Confederated International Conferences "On the Move to Meaningful Internet Systems"*, LNCS 4804, Springer 1567–1582
5. Albiero, V., King, M.C., Bowyer, K.W.: Gendered differences in face recognition accuracy explained by Hairstyles, Makeup, and Facial Morphology. *IEEE Transactions on Information Forensics and Security*, 17: 127–137 (2022)
6. Almeida,F., Xexeo F.: Word Embeddings: a survey. CoRR abs/1901.09069 (2019)
7. Alsaheel, A., Nan, Y., Ma, S., Yu, L., Walkup, G., Celik, Z.B., Zhang, X., Xu, D., 2021. ATLAS: A sequence-based learning approach for attack investigation. *Proceedings of the 30th USENIX Security Symposium (USENIX Security'21)*, 3005–3022
8. Al-Kasassbeh, M., Mohammed, S., Alauthman, M., Almomani, A.: Feature selection using a machine learning to classify a malware. *Handbook of Computer Networks and Cyber Security*, 889–904, 2020, Springer
9. Agrawal, R., Imieliński, T., Swami, A.: Mining association rules between sets of items in large databases. *Proceedings of the 1993 ACM SIGMOD International Conference on Management of Data (SIGMOD'93)*, 207–216
10. Agrawal, R., Srikant, R.: Fast algorithms for mining association rules in large databases. *Proceedings of 20th International Conference on Very Large Data Bases (VLDB'94)*, 487–499

E. Bertino et al., *Machine Learning Techniques for Cybersecurity*, Synthesis Lectures on Information Security, Privacy, and Trust,
https://doi.org/10.1007/978-3-031-28259-1

11. Alaiz-Moreton, H., Aveleira-Mata, J., Ondical-Garcia, J., Muñoz-Castañeda, A. L., García, I., Benavides, C.: Multiclass classification procedure for detecting attacks on MQTT-IoT protocol, *Complexity*, 2019

12. Aljabri, M., Alahmadi, AA., Mohammad, RM., Aboulnour, M., Alomari, DM., Almotiri, SH.: Classification of firewall log data Using multiclass machine learning models. *Electronics*, 11(12):1851, 2022

13. Allen, B., McGough, A.S., Devlin, M.: Toward a framework for teaching artificial intelligence to a higher education audience. *ACM Trans. Comput. Educ.*, 22(2): 15:1–15:29 (2022)

14. Allen, A.: Dismantling the "Black Opticon": privacy, race equity, and online data-protection reform. *The Yale Law Journal Forum*, 131, 2021–2022, https://www.yalelawjournal.org/forum/dismantling-the-black-opticon

15. Andow, B., Mahmud, S.Y., Whitaker, J., Enck, W., Reaves, B., Singh, K., Egelman, S., 2020. Actions speak louder than words:Entity-Sensitive privacy policy and data flow analysis with PoliCheck. *Proceedings of the 29th USENIX Security Symposium (USENIX Security'20)*, 985–1002

16. Andow, B., Mahmud, S.Y., Wang, W., Whitaker, J., Enck, W., Reaves, B., Singh, K., Xie, T., 2019. PolicyLint: Investigating Internal Privacy Policy Contradictions on Google Play. *Proceedings of the 28th USENIX Security Symposium (USENIX Security'19)*, 585–602

17. Aorato Labs: The Untold Story of the Target Attack Step by Step. August 2014. https://aroundcyber.files.wordpress.com/2014/09/aorato-target-report.pdf

18. Argento, L., Margheri, A., Paci, F., Sassone, V., Zannone, N.: Towards adaptive access control. *Proceedings of the 32nd IFIP Annual Conference on Data and Applications Security and Privacy (DBSec'18)*, 99–109, LNCS 10980, Springer

19. Ashok, M., Turner, M., Walsworth, R., Levine, E., Chandrakasan, A.: 2022. Hardware Trojan detection using unsupervised deep learning on quantum diamond microscope magnetic field images. *ACM Journal of Emerging Technologies in Computing Systems* 18(4): Article 67 (October 2022), https://doi.org/10.1145/3531010

20. Azad, M.A., Riaz, F., Aftab, A., Rizvi, S.K.J., Arshad, J., Atlam, H.F.: DEEPSEL: a novel feature selection for early identification of malware in mobile applications. *Future Generation Computer Systems*, 129:54–63 (2022)

21. Balakrishnan, G.,Reps, T.: Analyzing memory accesses in x86 executables. *Proceedings of the 13th International Conference on Compiler Construction (CC'04)*, 5–23, LNCS 2985, Springer

22. Bao, C., Forte, D., Srivastava, A.: On application of one-class SVM to reverse engineering-based hardware Trojan detection. *Proceedings of the 15th International Symposium on Quality Electronic Design (ISQED 2014)*, 47–54

23. Bao, C., Forte, D., Srivastava, A.: On reverse engineering-based hardware trojan detection. *IEEE Trans. Comput. Aided Des. Integr. Circuits Syst.*, 35(1): 49–57 (2016)

24. Barut, O., Luo, Y., Zhang, T., Li, W., Li, P.: Netml: A challenge for network traffic analytics, *arXiv preprint* arXiv:2004.13006 (2020)

25. Basumallik, S., Ma, R., Eftekharnejad, S.: Packet-data anomaly detection in PMU-based state estimator using convolutional neural network, *International Journal of Electrical Power & Energy Systems*, 690-702, 2019

26. Batini, C., Scannapieco, M.: Data and Information Quality - Dimensions, Principles and Techniques. Springer 2016

27. Baek, S., Jeon, J., Jeong, B., Jeong, Y. S.: Two-stage hybrid malware detection using deep learning. *Human-centric Computing and Information Sciences*, 11(27): 10-22967, 2021

28. Bertino, E., Jabal, A.A., Calo, S., Verma, D. and Williams, C.: The challenge of access control policies quality. *Journal of Data and Information Quality (JDIQ)*, 10(2):1–6 (2018)

29. Bertino, E., Merrill, S., Nesen, A., Utz, C.: Redefining data transparency: a multidimensional approach. *IEEE Computer*, 52(1): 16–26 (2019)

30. Bertino, E.: Attacks on artificial intelligence. *IEEE Security & Privacy*, 19(1):103–104 (2021)

31. Bertino, E.: The Persistent problem of software insecurity. *IEEE Security & Privacy*, 20(3):107–108 (2022)

32. Bodei, C., Degano, P., Galletta, L., Focardi, R., Tempesta, M., Veronese L.: Language-independent synthesis of firewall policies. *Proceedings of the 2018 IEEE European Symposium on Security and Privacy (EuroS&P'18)*, 92–106

33. Bonawitz, K., Eichner, H., Grieskamp, W., Huba, D., Ingerman, A., Ivanov, V., Kiddon, C., Konečný, J., Mazzocchi, S., McMahan, B., Van Overveldt, T.: Towards federated learning at scale: System design.*Proceedings of Machine Learning and Systems (MLSys'11)*, 1:374–88

34. Blei, D.M., Ng, A.Y., Jordan, M.I.: Latent dirichlet allocation. *Journal of machine Learning research*, 2003, 3 Jan, pp. 993–1022.

35. Breiman, L.: Random forests. *Machine learning.*, 45(1):5–32 (2001)

36. Breiman, L.: Some properties of splitting criteria. *Machine learning.* 24(1):41–7 (1996)

37. Bromley, J., Guyon, I., LeCun, Y., Sackinger, E., Shah, R.: Signature verification using a siamese time delay neural network. *Proceedings of the 7th Conference on Neural Information Processing Systems (NIPS'93)*, 737–744

38. Carrillo-Mondéjar, J., Martínez, J. L., Suarez-Tangil, G.: Characterizing Linux-based malware: findings and recent trends. *Future Generation Computer Systems*, 110:267–281 (2020)

39. Canizo, M., Triguero, I., Conde, A., Onieva, E.: Multi-head CNN–RNN for multi-time series anomaly detection: An industrial case study. *Neurocomputing*, 246–260 (2019)

40. Catal, C., Gunduz, H., Ozcan, A.: Malware detection based on graph attention networks for intelligent transportation systems. *Electronics*, 10(20), 2534 (2021)

41. Ceron, D.: AI, machine learning and deep learning: What's the difference? https://www.ibm.com/blogs/systems/ai-machine-learning-and-deep-learning-whats-the-difference/

42. Chakraborty, R.S., Narasimhan, S., Bhunia, S.: Hardware Trojan: threats and emerging solutions. *Proceedings of the 2009 IEEE International High Level Design Validation and Test Workshop (HLDVT'09)*, 166–171

43. Chaudhari, S., et al.: An attentive survey of attention models. *ACM Transactions on Intelligent Systems and Technology (TIST)*, 12.5: 1–32 (2021)

44. Chen, Y., Poskitt, C. M., Sun, J., Adepu, S., Zhang, F.: Learning-guided network fuzzing for testing cyber-physical system defences. *Proceedings of the 34th IEEE/ACM International Conference on Automated Software Engineering (ASE'19)*, 962–973

45. Chen, Y., Tang, D., Yao, Y., Zha, M., Wang, X., Liu, X., Tang, H., Zhao, D.: Seeing the forest for the trees: understanding security hazards in the 3GPP ecosystem through intelligent analysis on change requests. *Proceedings of the 31st USENIX Security Symposium (USENIX Security'22)*, 17–34

46. Chen, Y., Yao, Y., Wang, X., Xu, D., Yue, C., Liu, X., Chen, K., Tang, H., Liu, B.: Bookworm game: Automatic discovery of LTE vulnerabilities through documentation analysis. *Proceedings of the 42nd IEEE Symposium on Security and Privacy (S&P'21)*, 1197–1214

47. Chen, Y., Xing, L., Qin, Y., Liao, X., Wang, X., Chen, K., Zou, W., 2019. Devils in the guidance: predicting logic vulnerabilities in payment syndication services through automated documentation analysis. *Proceedings of the 28th USENIX Security Symposium (USENIX Security'19)*, 747–764

48. Chen, Y., Zha, M., Zhang, N., Xu, D., Zhao, Q., Feng, X., Yuan, K., Suya, F., Tian, Y., Chen, K., Wang, X.: Demystifying hidden privacy settings in mobile apps. *Proceedings of the 40th IEEE Symposium on Security and Privacy (S&P'19)*, 570–586

49. Cho, K., Bahdanau, D., Bougares., F., Schwenk., H., Bengio., Y.: Learning phrase representa-
    tions using RNN encoder-decoder for statistical machine translation. *Proceedings of the 2014
    Conference on Empirical Methods in Natural Language Processing (EMNLP'15)*, 1724–1734

50. Chua, Z.L., Shen, S., Saxena, P., Liang, Z., 2017. Neural nets can learn function type signatures
    from binaries. *Proceedings of the 26th USENIX Security Symposium (USENIX Security'17)*,
    99–116

51. Colantonio, A., Di Pietro, R., Ocello, A.: A cost-driven approach to role engineering. *Proceed-
    ings of the 23rd ACM Symposium on Applied Computing (SAC'08)*, 2129–2136

52. Collmann, J., Matei, S.: Ethical Reasoning in Big Data: An Exploratory Analysis. Springer,
    2016

53. Cotrini, C., Weghorn, T., Basin, D.: Mining ABAC rules from sparse logs. *Proceedings of the
    3rd IEEE European Symposium on Security and Privacy (EuroS&P'18)*, 31–46

54. Cozzi, E., Graziano, M., Fratantonio, Y., Balzarotti, D.: Understanding linux malware. *Pro-
    ceedings of the 39th IEEE Symposium on Security and Privacy (S&P'18)*, 161–175

55. CNN: Former SolarWinds CEO blames intern for 'solarwinds123' password leak. https://www.
    cnn.com/2021/02/26/politics/solarwinds123-password-intern/index.html

56. Cotrini, C., Corinzia, L., Weghorn, T., Basin, D.: The next 700 policy miners: A universal
    method for building policy miners. *Proceedings of the 26th ACM Conference on Computer and
    Communications Security (CCS'19)*, 95–112

57. Cowls, J., Floridi, L.: Prolegomena to a white paper on an ethical framework for a good AI
    society. June 19, 2018, Available at SSRN: https://ssrn.com/abstract=3198732 or http://dx.doi.
    org/10.2139/ssrn.3198732

58. Cunnington, D., Law, M., Russo, A., Lobo, J., Kaplan, L. M.: Towards neural-symbolic learning
    to support human-agent operations. *Proceedings of the 24th IEEE International Conference on
    Information Fusion (FUSION'21)*, 1–8

59. Chollet, F.: Xception: Deep learning with depthwise separable convolutions. *Proceedings of
    2017 the IEEE conference on computer vision and pattern recognition (CVPR'17)*, 1251–1258

60. Dai, Y., Li, H., Qian, Y., Lu, X. A malware classification method based on memory dump
    grayscale image. *Digital Investigation*, 27:30–37 (2018)

61. Datta, A., Sen, S., Zick, Y.: Algorithmic transparency via quantitative input influence: theory
    and experiments with learning systems. *Proceedings of the 37th IEEE Symposium on Security
    and Privacy (S&P'16)*, 598–617

62. Das S., Sural S., Vaidya J., Atluri V.: Using gini impurity to mine attribute-based access control
    policies with environment attributes. *Proceedings of the 23nd ACM on Symposium on Access
    Control Models and Technologies (SACMAT'18)*, 213–215

63. Devlin, J., Chang, M.W., Lee, K., Toutanova, K., 2018. Bert: Pre-training of deep bidirectional
    transformers for language understanding. *arXiv preprint* arXiv:1810.04805.

64. De Donno, M., Dragoni, N., Giaretta, A., Spognardi, A.: Analysis of DDoS-capable IoT mal-
    wares. *Proceedings of IEEE the Federated Conference on Computer Science and Information
    Systems (FedCSIS 2017)*, 807–816

65. Drucker, H., Burges, C. J. C., Kaufman, L., Smola, A. J., Vapnik, V.: Support vector regression
    machines. *Proceedings of the 9th conference on Advances in Neural Information Processing
    Systems (NIPS'96)*, 155–161

66. Du, M., Chen, Z., Liu, C., Oak, R., Song, D.: Lifelong anomaly detection through unlearn-
    ing. *Proceedings of the 26th ACM Conference on Computer and Communications Security
    (CCS'19)*, 1283–1297

67. Dutta, K., Krishnan, P., Mathew, M., Jawahar, C.V: Improving CNN-RNN hybrid networks for
    handwriting recognition. *Proceedings of the 16th IEEE International Conference on Frontiers
    in Handwriting Recognition (ICFHR'18)*, 80–85

68. Ertam, F., Kaya, M.: Classification of firewall log files with multiclass support vector machine. *Proceedings of the 6th IEEE International symposium on digital forensic and security (ISDFS'18)*, 1–4

69. Golnabi, K., Min, R. K., Khan, L., Al-Shaer, E.: Analysis of firewall policy rules using data mining techniques. *Proceedings of the 10th IEEE/IFIP Network Operations and Management Symposium (NOMS'06)*, 305–315

70. Drozdov, A., Law, M., Lobo, J., Russo, A., Don, M. W.: Online Symbolic Learning of Policies for Explainable Security. *Proceedings of the 3rd IEEE International Conference on Trust, Privacy and Security in Intelligent Systems and Applications (TPS-ISA'21)*, 269–278

71. Elhadi, A. A. E., Maarof, M. A., Barry, B. I.: Improving the detection of malware behaviour using simplified data dependent API call graph. *International Journal of Security and Its Applications*, 7(5):29–42 (2013)

72. Elsersy, W. F., Feizollah, A., Anuar, N. B.: The rise of obfuscated Android malware and impacts on detection methods. *PeerJ Computer Science*, 8, e907

73. El Merabet, H., Hajraoui, A.: A survey of malware detection techniques based on machine learning. *International Journal of Advanced Computer Science and Applications*, 10(1) (2019), http://dx.doi.org/10.14569/IJACSA.2019.0100148

74. Emanuelsson, P.: A Comparative study of industrial static analysis tools. *Electronic Notes in Theoretical Computer Science*, 217:5–21 (2008)

75. 1st International Workshop on Ethics in Computer Security (EthiCS 2022). June 10, 2022, https://ethics-workshop.github.io/2022/

76. Fernandes, G., Rodrigues, J. J., Carvalho, L. F., Al-Muhtadi, J. F., & Proença, M. L.: A comprehensive survey on network anomaly detection. *Telecommunication Systems*, 70(3): 447–489 (2019)

77. Fayazbakhsh, S.K., Chiang, L., Sekar, V., Yu, M., Mogul, J. C.: Enforcing network-wide policies in the presence of dynamic middlebox actions using FlowTags. *Proceedings of the 11th USENIX Symposium on Networked Systems Design and Implementation (NSDI'14)*, 543–546

78. Feng, C., Li, T., Chana, D.: Multi-level anomaly detection in industrial control systems via package signatures and LSTM networks. *Proceedings of the 47th Annual IEEE/IFIP International Conference on Dependable Systems and Networks (DSN'17)*, 261–272

79. Fioraldi, A., Maier, D., Eißfeldt, H., Heuse, M.: AFL++: Combining incremental steps of fuzzing research. *14th USENIX Workshop on Offensive Technologies (WOOT'20)*

80. Fisler, K., Krishnamurthi, S., Meyerovich, L. A., Tschantz, M. C.: Verification and change-impact analysis of access-control policies. *Proceedings of the 27th international conference on Software engineering (ICSE'05)*, 196–205

81. Gad, M., Aboelmaged, M., Mashaly, M., el Ghany, M. A. A.: Efficient sequence generation for hardware verification using machine learning. *Proceedings of the 28th IEEE International Conference on Electronics, Circuits, and Systems (ICECS 2021)*, 1–5, https://doi.org/10.1109/ICECS53924.2021.9665495

82. Gade, K., Geyik, S., Kenthapadi, K., Mithal, V., Taly, A.: Explainable AI in industry. *Proceedings of the 25th ACM SIGKDD International Conference on Knowledge Discovery & Data Mining (KDD'19)*, 3203–3204

83. Garcia, J., Fernandez, F.: A comprehensive survey on safe reinforcement learning. *Journal of Machine Learning Research*, 16(1):1437–1480 (2015)

84. Gaur, P., Rout, S. S., Deb, S.: Efficient hardware verification using machine learning approach. *Proceedings of the IEEE International Symposium on Smart Electronic Systems (iSES 2019) (Formerly iNiS)*, 168–171

85. Gazzillo, P., Cohen, M.B.: Bringing together configuration research: towards a common ground. *Proceedings of the 2022 ACM SIGPLAN International Symposium on New Ideas, New Paradigms, and Reflections on Programming and Software (Onward! '22)*, 11 pages

86. Gilpin, L. H., Bau, D., Yuan, B. Z., Bajwa, A., Specter, M., Kagal, L.: Explaining explanations: An overview of interpretability of machine learning. *Proceedings of the IEEE 5th International Conference on Data Science and Advanced Analytics (DSAA 2018)*, 80–89

87. Golovin, D., Solnik, B., Moitra, S., Kochanski, G., Elliot, J., Sculley, D.: Google Vizier: A Service for black-box optimization. *Proceedings of the 23rd ACM SIGKDD International Conference on Knowledge Discovery and Data Mining (SIGKDD'17)*, 1487–1495

88. Goodfellow, I., Pouget-Abadie, J., Mirza, M., Xu, B., Warde-Farley, D., Ozair, S., Courville, A. & Bengio, Y. Generative adversarial nets. *Advances In Neural Information Processing Systems*. **27** (2014)

89. Goodfellow, I., Bengio, Y., Courville, A. Deep learning. MIT press, 2016

90. Google: Syzkaller. https://github.com/google/syzkaller

91. Goyal, P., Ferrara, E.: Graph embedding techniques, applications, and performance: a survey. arXiv:1705.02801

92. Goutte, C., Toft, P., Rostrup, E., Nielsen, FÅ., Hansen, L. K.: On clustering fMRI time series. *NeuroImage*. 1999 Mar 1;9(3):298–310.

93. Greengard, S.: The worsening state of ransomware. *Commununications of ACM* 64(4): 15–17 (2021)

94. Guidotti, R.: Counterfactual explanations and how to find them: literature review and benchmarking. *Data Mining and Knowledge Discovery*, 2022, https://doi.org/10.1007/s10618-022-00831-6

95. Han, R., Yang, C., Ma, S., Ma, J., Sun, C., Li, J., Bertino, E.: Control parameters considered harmful: detecting range specification bugs in drone configuration modules via learning-guided search. *Proceedings of the 44th IEEE International Conference on Software Engineering (ICSE'22)*, 462–473

96. Harter, G. T., Rowe, N. C.: Testing Detection of K-Ary Code Obfuscated by Metamorphic and Polymorphic Techniques. *Proceedings of National Cyber Summit*, 110–123, 2021

97. Hutter, F., Kotthoff, L., Vanschoren, J.: Automated machine learning: methods, systems, challenges, *Springer Nature* (2019)

98. He, H., Ji, Huang, H.H. Illuminati: towards explaining graph neural networks for cybersecurity analysis. *Proceedings of the 7th IEEE European Symposium on Security and Privacy (EuroS&P'22)*, 74–89

99. He, K., Zhang, X., Ren, S., Sun, J.: Deep residual learning for image recognition. *Proceedings of the 2016 IEEE Conference on Computer Vision and Pattern Recognition (CVPR 2016)*, 770–778

100. Hospodar, G., Gierlichs, B., De Mulder, E., Verbauwhede, I., Vandewalle, J.: Machine learning in side-channel analysis: a first study. *Journal of Cryptographic Engineering*, 1(4):293–302 (2011)

101. Hospodar, G., Maes, R., Verbauwhede, I.: Machine learning attacks on 65mm Arbiter PUFs: Accurate modeling poses strict bounds on usability. *Proceedings of the 4th IEEE International Workshop on Information Forensics and Security (WIFS'12)*, 37–42

102. Hu, V. C., Ferraiolo, D., Kuhn, R., Friedman, A., Lang, A., Cogdell, M., Schnitzer, A., Sandlin, K., Miller, R., Scarfone, K.: Guide to attribute based access control (abac) definition and considerations (draft). *NIST special publication 800*, no. 162 (2013): 1–54

103. Hu, Y., Kuang, W., Gao, Y., Li, K., Li, W., Qin, Z., Kenli, L., Zhang, J.: Artificial intelligence security: threats and countermeasures. *ACM Computing Surveys*, 55(1):1–36 (2023)

104. Huang, X., Ma, L., Yang, W., Zhong, Y.: A method for windows malware detection based on deep learning. *Journal of Signal Processing Systems*, 93(2):265–273 (2021)

105. Hu, W., Le, R., Liu, B., Ji, F., Ma, J., Zhao, D. Yan, R.: Predictive adversarial learning from positive and unlabeled data. *Proceedings of the 35th AAAI Conference on Artificial Intelligence (AAAI'21)*, 7806–7814

106. Huang, G., Liu, Z., Van Der Maaten, L., Weinberger, KQ.: Densely connected convolutional networks. *Proceedings of the 2017 IEEE Conference on Computer Vision and Pattern Recognition (CVPR'17)*, 4700–4708

107. Huang, Z.: Extensions to the k-means algorithm for clustering large data sets with categorical values. *Data mining and knowledge discovery*, 1998 Sep;2(3):283–304.

108. Huang, Z., Wang, Q., Chen, Y., Jiang, X.: A survey on machine Learning against hardware Trojan attacks: Recent Advances and Challenges. *IEEE Access*, 8: 10796–10826 (2020)

109. Hamon, R., Junklewitz, H., Sanchez, I.: *Robustness and explainability of Artificial Intelligence - From technical to policy solutions* EUR 30040, Publications Office of the European Union, Luxembourg, Luxembourg, 2020, ISBN 978-92-79-14660-5 (online), https://doi.org/10.2760/57493 (online), JRC119336

110. Holland, J., Schmitt, P., Feamster, N., Mittal, P.: New directions in automated traffic analysis. *Proceedings of the 28th ACM Conference on Computer and Communications Security (CCS'21)*, 3366–3388

111. Hundman, K., Constantinou, V., Laporte, C., Colwell, I., Soderstrom, T.: Detecting spacecraft anomalies using lstms and nonparametric dynamic thresholding, *Proceedings of the 24th ACM SIGKDD International Conference on Knowledge Discovery & Data Mining (KDD'18)*, 387–395

112. Irshad, M., Al Khateeb, H. M., Mansour, A., Ashawa, M., & Hamisu, M.: Effective methods to detect metamorphic malware: a systematic review. *International Journal of Electronic Security and Digital Forensics*, 10(2):138–154 (2018)

113. Institute for Security and Technology: Combating Ransomware A Comprehensive Framework for Action: Key Recommendations from the Ransomware Task Force. April 2021, https://securityandtechnology.org/ransomwaretaskforce/report/

114. Ismail, K. A., Ghany, M. A.: Survey on machine learning algorithms enhancing the functional verification process. *Electronics*, 10(21), 2688 (2021)

115. Ismail, K. A., Ghany, M. A.: High performance machine learning models for functional verification of hardware designs. *Proceedings of the 3rd Novel Intelligent and Leading Emerging Sciences Conference (NILES'21)*, 15–18

116. Kravchik, M., Shabtai, A.: Detecting cyber attacks in industrial control systems using convolutional neural networks, *Proceedings of the 2018 Workshop on Cyber-Physical Systems Security and Privacy*, 72–83, 2018

117. LeCun, Y., Bengio, Y. & Hinton, G. Deep learning. *Nature* 521: 436-444 (2015)

118. Leef, S.: Automatic implementation of secure silicon. *Proceedings of the 29th Great Lakes Symposium on VLSI (GLSVLSI 2019)*, 3

119. Lyu, Y., Fang, Y., Zhang, Y., Sun, Q., Ma, S., Bertino, E., Lu, K., Li, J.: Goshawk: hunting memory corruptions via structure-aware and object-centric memory operation synopsis. *Proceedings of the 43rd IEEE Symposium on Security & Privacy (S&P'22)*, 2096–2133

120. Li, W., Wasson, Z., Seshia, S.A.: Reverse engineering circuits using behavioral pattern mining. *Proceedings of the 2012 IEEE Conference on Hardware-Oriented Security and Trust (HOST'12)*, 83–88

121. Liang, J., Guo, W., Luo, T., Hanovar, V., Wang, G., Xing, X.: FARE: enabling fine-grained attack categorization under low-quality labeled data. *Proceedings of the 28th Network and Distributed System Security Symposium (NDSS'21)*

122. Jabal, A.A., Bertino, E., Lobo, J., Law, M., Russo, A., Calo, S., Verma, D.: Polisma - a frame-work for learning attribute-based access control policies. *Proceedings of the 25th European Symposium on Research in Computer Security (ESORICS 2020), Part I*, LNCS 12308, Springer, 523–544

123. Jabal, A.A., Bertino, E., Lobo, J., Verma, D., Calo, S., Russo, A.: FLAP–A Federated Learning Framework for Attribute-based Access Control Policies. *arXiv preprint* arXiv:2010.09767, 2020 Oct. 19

124. Jeffcock, P.: What's the difference between AI, machine learning, and deep learning? https://blogs.oracle.com/bigdata/post/whatx27s-the-difference-between-ai-machine-learning-and-deep-learning

125. Jaiswal, A., Ramesh Babu, A., Zaki Zadeh, M., Banerjee, D., Makedon, F.: A survey on contrastive self-supervised learning. *Technologies* 9:2 (2021), https://dx.doi.org/10.3390/technologies9010002

126. Javed, A. R., Ahmed, W., Alazab, M., Jalil, Z., Kifayat, K., Gadekallu, T. R.: A comprehensive survey on computer forensics: state-of-the-art, tools, techniques, challenges, and future directions. *IEEE Access* 10: 11065–11089 (2022)

127. Jain, A. K., Dubes, R. C.: Algorithms for clustering data. *Prentice-Hall, Inc.*, 1988 Jul 1.

128. Jing, L., Tian, Y.: Self-supervised visual feature learning with deep neural networks: a Survey. *IEEE Trans. Pattern Anal. Mach. Intell.* 43(11): 4037–4058 (2021)

129. Kambar, M. E. Z. N., Esmaeilzadeh, A., Kim, Y., Taghva, K.: A survey on mobile malware detection methods using machine learning. *Proceedings of the 12th IEEE Annual Computing and Communication Workshop and Conference (CCWC'22)*, 0215–0221

130. Karimi, L., Abdelhakim, M., Joshi, J.: Adaptive ABAC policy learning: a reinforcement learning approach. *arXiv preprint* arXiv:2105.08587, 2021 May 18

131. Karimi, L., Joshi, J.: An unsupervised learning based approach for mining attribute based access control policies. In *2018 IEEE International Conference on Big Data (Big Data)*, 2018 Dec 10 (pp. 1427–1436). IEEE.

132. Karita, S., Chen, N., Hayashi, T., Hori, T., Inaguma, H., Jiang, Z., Someki, M., Soplin, N.E., Yamamoto, R., Wang, X., Watanabe, S.: A comparative study on transformer vs rnn in speech applications. *Proceedings of the 2019 IEEE Automatic Speech Recognition and Understanding Workshop (ASRU)*, 449–456

133. Katsis, C., Cicala, F., Thomsen, D., Ringo, N., Bertino, E.: NEUTRON: A graph-based pipeline for zero-trust network architectures. *Proceedings of the 12th ACM Conference on Data and Application Security and Privacy (CODASPY'22)*, 167–178

134. Kaelbling, L.P., Littman, M. L., Moore, A. W.: Reinforcement learning: a survey. *Journal of Artificial Intelligence Research (JAIR)* 4:237–285 (1996)

135. Kieu, T., Yang, B., Jensen, C.: Outlier detection for multidimensional time series using deep neural networks. *Proceedings of the 19th IEEE International Conference on Mobile Data Management (MDM'18)*, 125–134

136. Kim, K., Jeong, D. R., Kim, C. H., Jang, Y., Shin, I., Lee, B.: HFL: Hybrid fuzzing on the Linux kernel. *Proceedings of the 27th Network and Distributed System Security Symposium (NDSS'20)*

137. Krose, B., Smagt, P.: An introduction to neural networks. 2011, http://14.99.188.242:8080/jspui/bitstream/123456789/1991/1/An%20Introduction%20to%20Neural%20Networks.pdf

138. Kudo, T.: Subword regularization: improving neural network translation models with multiple subword candidates. *Proceedings of the Proceedings of the 56th Annual Meeting of the Association for Computational Linguistics (ACL'18)*, 66–75

139. Lashkari, A.H., Kadir, A.F.A., Taheri, L., Ghorbani, A.A.,: Toward developing a systematic approach to generate benchmark android malware datasets and classification. *Proceedings of the 2018 IEEE International Carnahan Conference on Security Technology (ICCST'18)*, 1–7

140. Law, M., Russo, A., Bertino, E., Broda, K., Lobo, J.: FastLAS: scalable inductive logic programming incorporating domain-specific optimisation Criteria. *Proceedings of the 34th AAAI Conference on Artificial Intelligence (AAAI'20)*, 2877–2885

141. Law, M., Russo, A., Broda, K., Bertino, E.: Scalable non-observational predicate learning in ASP. *Proceedings of the 38th International Joint Conference on Artificial Intelligence (IJCAI'21)*, 1936–1943

142. Lee, H., Mudgerikar, A., Kundu, A., Li, N., Bertino, E.: An Infection-Identifying and Self-Evolving System for IoT Early Defense from Multi-Step Attacks. *Proceedings of the 27th European Symposium on Research in Computer Security (ESORICS'22)*, Springer LNCS 13555, 549–568

143. Levine, S., Kumar, A., Tucker, G., Fu, F.: Offline reinforcement learning: tutorial, review, and perspectives on open problems. *CoRR abs/2005.01643 (2020)*

144. Levine, S.: The case for real-world reinforcement learning. https://www.youtube.com/watch?v=Ik1nS2E4ar4 Talk recorded for the Deep Reinforcement Learning Workshop at NeurIPS 2021. Accessed on June 1, 2022

145. Li, Y.: Deep reinforcement learning. *CoRR abs/1810.06339 (2018)*

146. Li, Y.: Deep reinforcement learning: opportunities and challenges. *CoRR abs/2202.11296 (2022)*

147. Li, D., Chen, D., Jin, B., Shi, L., Goh, J., Ng, S.: MAD-GAN: Multivariate anomaly detection for time series data with generative adversarial networks, *Proceedings of the International Conference on Artificial Neural Networks (ICANN'19)*, 703–716.

148. Li, C., Li, F., Hao, Z., Yin, L., Sun, Z., Wang, C.: An IoT Crossdomain Access Decision-Making Method Based on Federated Learning. *Wireless Communications and Mobile Computing,* Volume 2021, Article ID 8005769, https://doi.org/10.1155/2021/8005769

149. Luo, Y., Xiao, Y., Cheng, L., Peng, G., Yao, D.: Deep learning-based anomaly detection in cyber-physical systems: Progress and opportunities. *ACM Computing Surveys (CSUR)*, 54(5): 106:1–106:36 (2021)

150. Lyu, Y., Fang, Y., Zhang, Y., Sun, Q., Ma, S., Bertino, E., Lu, K., Li, J.: Goshawk: hunting memory corruptions via structure-aware and object-centric memory operation synopsis. *Proceedings of the 43rd IEEE Symposium on Security and Privacy (S&P'22)*, 2096–2113

151. Ma, S., Bertino, E., Nepal, S., Li, J., Ostry, D., Deng, R.H., Jha, S.: Finding flaws from password authentication code in android apps. *Proceedings of the 24th European Symposium on Research in Computer Security (ESORICS'19)*, Springer LNCS 11735, 619–637

152. Manès, V. J. M., Han, H., Han, C., Cha, S. K., Egele, M., Schwartz, E. J., Woo, M.: The art, science, and engineering of fuzzing: a survey. *IEEE Transactions on Software Engineering*, 47(11): 2312–2331 (2021)

153. Mehdi, B., Ahmed, F., Khayyam, S. A., Farooq, M.: Towards a theory of generalizing system call representation for in-execution malware detection. *Proceedings of the 18th IEEE International Conference on Communications (ICC'11)*, 1–6

154. Mehnaz, S., Dibbo, S.V., Kabir, E., Li, N., Bertino, E.: Are your sensitive attributes private? novel model inversion attribute inference attacks on classification models. *Proceedings of the 31st USENIX Security Symposium (USENIX Security'22)*, 4579–4596

155. Mehnaz, S., Mudgerikar, A., Bertino, E.: RWGuard: a real-time detection system against cryptographic ransomware. *Proceedings of the 21st International Symposium (RAID 2018)*, Springer LNCS 11050, 114–136

156. Mehrabi, N., Morstatter, F., Saxena, N., Lerman, K., Galstyan, A.: A survey on bias and fairness in machine learning. *ACM Computing Surveys*, 54(6):1–35 (2021)

157. Mendsaikhan, O., Hasegawa, H., Yamaguchi, Y., Shimada, H.: Identification of cybersecurity specific content using the Doc2Vec language model. *Proceedings of the IEEE 43rd annual computer software and applications conference (COMPSAC'19)*, 396–401

158. Mirsky, Y., Doitshman, T., Elovici, Y., Shabtai, A.: Kitsune: An Ensemble of Autoencoders for Online Network Intrusion Detection. *Proceedings of the 25th Network and Distributed System Security (NDSS'18)*

159. Molloy, I., Park Y., Chari, S.: Generative models for access control policies: applications to role mining over logs with attribution. In *Proceedings of the 17th ACM Symposium on Access Control Models and Technologies*, 2012 Jun 20 (pp. 45–56).

160. Moussaileb, R., Cuppens, C., Lanet, J.-L., Le Bouder, H.: A survey on windows-based ransomware taxonomy and detection mechanisms. *ACM Computing Surveys*, 54(6): 117:1–117:36 (2021)

161. Mudgerikar, A., Bertino, E.: Jarvis: moving towards a smarter internet of things. *Proceedings of the 40th IEEE International Conference on Distributed Computing Systems (ICDCS'20)*, 122–134

162. Mudgerikar, A., Bertino, E., Lobo, D., Lobo, J.: A Security-constrained reinforcement learning framework for software defined networks. *Proceedings of the 31st IEEE International Conference on Communications (ICC'21)*, 1–6

163. Mudgerikar, A., Sharma, P., Bertino, E.: E-spion: A system-level intrusion detection system for IoT devices. *Proceedings of the 14th ACM Asia conference on computer and communications security (ASIACCS'19)*, 493–500

164. Nakhodchi, S., Upadhyay, A., Dehghantanha, A.: A comparison between different machine learning models for iot malware detection. *Security of Cyber-Physical Systems*, 195–202, 2020, Springer

165. Napiah, M. N., Idris, M. Y. I. B., Ramli, R., Ahmedy, I.: Compression Header Analyzer Intrusion Detection System (CHA - IDS) for 6LoWPAN Communication Protocol, *IEEE Access*, 6:16623–16638 (2018)

166. Nataraj, L., Karthikeyan, S., Jacob, G., Manjunath, B. S.: Malware images: visualization and automatic classification. *Proceedings of the 8th International Symposium on Visualization for Cybersecurity (VizSec'11,)*, 1–7

167. Ndichu, S., Kim, S., Ozawa, S., Misu, T., & Makishima, K.: A machine learning approach to detection of JavaScript-based attacks using AST features and paragraph vectors. *Applied Soft Computing*, 84, 105721, 2019.

168. Ni, Q., Lobo, J., Calo, S., Rohatgi, P., Bertino, E.: Automating role-based provisioning by learning from examples. *Proceedings of the 14th ACM Symposium on Access Control Models and Technologies, (SACMAT'09)*, 75–84

169. Ng, A.: Why is deep learning taking off? https://youtu.be/xflCLdJh0n0

170. Nguyen, A.: Papers on explainable artificial intelligence. GitHub Repository https://github.com/anguyen8/XAI-papers

171. Nguyen, T. D., Marchal, S., Miettinen, M., Fereidooni, H., Asokan, N., Sadeghi, A.: DÏoT: A federated self-learning anomaly detection system for IoT, *Proceedings of the IEEE 39th International Conference on Distributed Computing Systems (ICDCS'19)*, 756–767

172. Nobi, MN., Krishnan, R., Huang, Y., Shakarami, M., Sandhu, R.: Toward deep learning based access control. *Proceedings of the 12th ACM Conference on Data and Application Security and Privacy (CODASPY'22)*, 143–154

173. Nowroozi, E., Dehghantanha, A., Parizi, R.M., Choo, R.: A survey of machine learning techniques in adversarial image forensics. CoRR abs/2010.09680 (2020)

174. Otoum, Y., Liu, D., Nayak, A.: DL-IDS: a deep learning–based intrusion detection framework for securing IoT. *Transactions on Emerging Telecommunications Technologies*, 2019

175. Pacheco, M.L., von Hippel, M., Weintraub, B., Goldwasser, D., Nita-Rotaru, C.: Automated attack synthesis by extracting finite state machines from protocol specification documents. *Proceedings of the 43rd IEEE Symposium on Security and Privacy (S&P'22)*, 51–68

176. Pan, S. J., Yang, Q.: A survey on transfer learning. *IEEE Transactions on knowledge and data engineering*, 22(10): 1345–1359 (2009)

177. Pasikhani, A. M., Clark, A. J., Gope, P.: Reinforcement-learning-based IDS for 6LoWPAN, *Proceedings of the 20th IEEE International Conference on Trust, Security, and Privacy in Computing and Communications (TRustCom'21)*, 1049–1060

178. Pazzanese, C.: Ethical concerns mount as AI takes bigger decision-making role in more industries. *The Harvard Gazette*, October 2020, https://news.harvard.edu/gazette/story/2020/10/ethical-concerns-mount-as-ai-takes-bigger-decision-making-role/

179. Pei, K., Guan, J., Williams-King, D., Yang, J., Jana, S.: XDA: Accurate, robust disassembly with transfer learning. *Proceedings of the 28th Network and Distributed System Security Symposium (NDSS'21)*

180. Pendleton, M., Garcia-Lebron, R., Cho, J.-H., Xu, S.: A Survey on systems security metrics. *ACM Computing Surveys*, 49(4): 62:1–62:35 (2017)

181. Pes, B.: Ensemble feature selection for high-dimensional data: a stability analysis across multiple domains. *Neural Computing and Applications*, 32(10): 5951–5973 (2020)

182. Sayyad Shirabad, J., Menzies, T.J.: The PROMISE Repository of Software Engineering Databases. School of Information Technology and Engineering, University of Ottawa, Canada, 2005. http://promise.site.uottawa.ca/SERepository

183. Quinlan, J.R.: Generating production rules from decision trees. *Proceedings of the 10th International Joint Conference on Artificial Intelligence (IJCAI'87)*, 304–307

184. Rabadi, D., Teo, S. G.: Advanced windows methods on malware detection and classification. *Proceedings of the 36th Annual Computer Security Applications Conference (ACSAC'20)*, 54–68

185. Rajpal, M., Blum, W., Singh, R.: Not all bytes are equal: Neural byte sieve for fuzzing. Microsoft Research. ArXiv, 2017.

186. Raulin, V., Gimenez, P. F., Han, Y., Tong, V. V. T.: Towards a Representation of Malware Execution Traces for Experts and Machine Learning. *RESSI 2022-Meeting place for Research and Education in Information Systems Security*, 2022.

187. Reddy, S., Lemieux, C., Padhye, R., Sen, K.: . Quickly generating diverse valid test inputs with reinforcement learning. *Proceedings of the 42nd International Conference on Software Engineering (ICSE'20)*, 1410–1421

188. Regazzoni, F., Bhasin, S., Pour, A.A., Alshaer, I., Aydin, F., Aydin, A., Beroulle, V., Di Natale, G., Franzon, P., Hely, D., Homma, N., Ito, A., Dirmanto]., Kashyap, P., Polian, I., Potluri, S., Ueno, R., Vatajelu, E.I,, Yli-Mayry, V.: Machine learning and hardware security: challenges and opportunities. *Proceedings of the 4th IEEE/ACM International Conference On Computer Aided Design (ICCAD'20)*, 141:1–141:6

189. Richardson, B., Gilbert, J.E.: A framework for fairness: a systematic review of existing fair AI solutions. https://arxiv.org/abs/2112.05700 (2021)

190. Rizvi, S. K. J., Aslam, W., Shahzad, M., Saleem, S., & Fraz, M. M.: PROUD-MAL: static analysis-based progressive framework for deep unsupervised malware classification of windows portable executable. *Complex & Intelligent Systems*, 8(1): 673–685 (2022)

191. Rocca, J.: Understanding variational autoencoders (VAEs). *Towards Data Science*, https://towardsdatascience.com/understanding-variational-autoencoders-vaes-f70510919f73, September 2019

192. Rose, S., Borchert, O., Mitchell, S., Connelly, S.: Zero trust architecture. Technical Report. *National Institute of Standards and Technology*, 2019

193. Rosen-Zvi, M., Chemudugunta, C., Griffiths, T., Smyth, P., Steyvers, M.: Learning author-topic models from text corpora. *ACM Transactions on Information Systems (TOIS)*, 2010, Jan 29;28(1):1–38.

194. Rosen-Zvi, M., Griffiths, T., Steyvers, M., Smyth, P.: The author-topic model for authors and documents. *arXiv preprint* arXiv:1207.4169, 2012 Jul 11.

195. Roussev, V.: Digital forensic science: issues, methods, and challenges. *Synthesis Lectures on Information Security, Privacy, & Trust*, Morgan & Claypool Publishers 2016

196. Rusak, G., Al-Dujaili, A., O'Reilly, U. M.: Ast-based deep learning for detecting malicious powershell. In *Proceedings of the 25th ACM SIGSAC Conference on Computer and Communications Security (CCS'18)*, 2276–2278

197. Sanders, M. W., Yue, C.: Mining least privilege attribute based access control policies. *Proceedings of the 35th Annual Computer Security Applications Conference (ACSAC'19)*, 404–416

198. Samarati, P., Vimercati, S. C.: Access control: Policies, models, and mechanisms. In *International School on Foundations of Security Analysis and Design*, 2000 Sep 18 (pp. 137–196), Springer

199. Sandhu, R. S.: Role-based access control. *Advances in computers* 46:287–295 (1998)

200. Sentanoe, S., Taubmann, B., Reiser, H. P. : Virtual machine introspection based SSH honeypot. *Proceedings of the 4th Workshop on Security in Highly Connected IT Systems*, 13–18, 2017

201. Shahzad, F., Javed, A.R., Jalil, Z., Iqbal, F.: Cyber forensics with machine earning. *Phung, D., Webb, G.I., Sammut, C. (eds) Encyclopedia of Machine Learning and Data Science*, Springer, https://doi.org/10.1007/978-1-4899-7502-7_987-1

202. She, D., Pei, K., Epstein, D., Yang, J., Ray, B., Jana, S.: NEUZZ: Efficient Fuzzing with Neural Program Smoothing. *Proceedings of the 40th IEEE Symposium on Security and Privacy (S&P'19)*, 803–817

203. Shukla, P.: ML-IDS: A machine learning approach to detect wormhole attacks in Internet of Things, *SAI 2017 Intelligent Systems Conference (IntelliSys)*, 234–240

204. Shin, E. C. R., Song, D., Moazzezi, R., 2015. Recognizing functions in binaries with neural networks. *Proceedings of the 24th USENIX Security Symposium (USENIX Security'15)*, 611–626

205. Smith, G.: What does "fairness" mean for machine learning systems? https://haas.berkeley.edu/wp-content/uploads/What-is-fairness_-EGAL2.pdf

206. Saha, S., Jap, D., Patranabis, S., Mukhopadhyay, D., Bhasin, S., Dasgupta, P.: Automatic characterization of exploitable faults: A machine learning approach. *IEEE Transactions on Information Forensics and Security*, 14(4):954–968 (2018)

207. Selvaganapathy, S., Sadasivam, S., & Ravi, V.: A review on android malware: attacks, counter-measures, and challenges ahead. *Journal of Cyber Security and Mobility*, 177–230 (2021)

208. Shao, S., McAleer, S., Yan, R., Baldi, P.: Highly accurate machine fault diagnosis using deep transfer learning. *IEEE Transactions on Industrial Informatics*, 15(4):2446–2455 (2018)

209. Sharma, T., Kechagia, M., Georgiou, S., Tiwari, R., Sarro, F.: A Survey on Machine Learning Techniques for Source Code Analysis. ArXiv, 2021

210. Shewalkar, A.: Performance evaluation of deep neural networks applied to speech recognition: RNN, LSTM and GRU. *Journal of Artificial Intelligence and Soft Computing Research*, 9(4):235–45 (2019)

211. Shin, H.: Case study: Real-world machine learning application for hardware failure detection. *Proceedings of the 18th Python in Science Conference (SCIPY 2019)*, 1–5

212. Shorten, C., Khoshgoftaar, TM.: A survey on image data augmentation for deep learning. *Journal of Big Data*. 6(1):1–48 (2019)

213. Singla, A., Bertino, E.: How deep learning is making information security more intelligent. *IEEE Security & Privacy*, 17(3):56–65 (2019)
214. Singla, A., Bertino, E., Verma, D.: Overcoming the lack of labeled data: training intrusion detection models using transfer learning. *Proceedings of the 5th IEEE International Conference on Smart Computing (SMARTCOMP'19)*, 69–74
215. Singla, A., Bertino, E., Verma, D.: Preparing network intrusion detection deep learning models with minimal data using adversarial domain adaptation. *Proceedings of the 15th ACM Asia Conference on Computer and Communications Security (ASIACCS'20)*, 127–140
216. Spinosa, E. J., de Carvalho, A. P., Gama, J.: Novelty detection with application to data streams. *Intelligent Data Analysis.* 13(3):405–22 (2009)
217. Studiawan, H., Sohel, F., Payne, C.: Anomaly detection in operating system logs with deep learning-based sentiment analysis. *IEEE Transactions on Dependable and Secure Computing*, 18(5):2136–2148.
218. Szandala, T.: Review and comparison of commonly used activation functions for deep neural networks. https://arxiv.org/abs/2010.09458
219. Soydaner, D. Attention mechanism in neural networks: where it comes and where it goes. *Neural Computing and Applications*, 34(16): 13371–13385 (2022)
220. Sutton, R.S., Barto, A.G.: Reinforcement learning: an introduction. MIT press, 2018
221. Taddeo, M.: The limits of deterrence theory in cyberspace. *Philosophy & Technology*, 31:339–355 (2018)
222. Tahir, R.: A study on malware and malware detection techniques. *International Journal of Education and Management Engineering*, 8(2), 20 (2018)
223. Tan, B., Karri, R.: Challenges and new directions for AI and hardware security. *Proceedings of the 63rd IEEE International Midwest Symposium on Circuits and Systems (MWSCAS'20)*, 277–280
224. Tang, R., Yang, Z., Li, Z., Meng, W., Wang, H., Li, Q., Sun, Y., Pei, D., Wei, T., Xu, Y., Liu., Y: ZeroWall: Detecting Zero-Day Web Attacks through Encoder-Decoder Recurrent Neural Networks. *Proceedings of the 2020 International Conference on Computer Communications (INFOCOM'20)*, 2479–2488
225. Tariq, S., Lee, S., Shin, Y., Lee, M., Jung, O., Chung, D., Woo, S.: Detecting anomalies in space using multivariate convolutional LSTM with mixtures of probabilistic PCA, *Proceedings of the 25th ACM SIGKDD international conference on knowledge discovery & data mining (KDD'19)*, 2123–2133
226. Tekiner, E., Acar, A., Uluagac, S., Engin Kirda, E., Selçuk. A.A.: SoK: Cryptojacking malware. *Proceedings of the 6th IEEE European Symposium on Security and Privacy (EuroS&P'21)* 120–139
227. Thomson, P:. Static analysis: an introduction: the fundamental challenge of software engineering is one of complexity. *ACM Queue*, 19(4): 29–41 (2021)
228. Thomson, P:. Static analysis. *Communications of ACM*, 65(1): 50–54 (2022)
229. Tongaonkar, A., Inamdar, N., Sekar, R.: Inferring Higher Level Policies from Firewall Rules. In *Proceedings of the 21st Large Installation System Administration Conference (LISA '07)*, 2007 Nov 11 (Vol. 7, pp. 1–10).
230. Tongaonkar, A. S. Fast pattern-matching techniques for packet filtering (Doctoral dissertation, State University of New York at Stony Brook), 2004.
231. Ucci, D., Aniello, L., Baldoni, R.: Survey of machine learning techniques for malware analysis. *Computer & Security*, 81: 123–147 (2019)
232. Van Ede, T., Aghakhani, H., Spahn, N., Bortolameotti, R., Cova, M., Continella, A., Van Steen, M., Peter, A., Kruegel, C., Vigna, G.: DEEPCASE: Semi-supervised contextual analysis of

security events. *Proceedings of the 43rd IEEE Symposium on Security and Privacy (S&P'22),* 522–539

233. Van Wyk, F., Wang, Y., Khojandi, A., Masoud, N.: Real-time sensor anomaly detection and identification in automated vehicles, *IEEE Transactions on Intelligent Transportation Systems,* 1264–1276 (2019)

234. Vasudevan, S., Jiang, W., Bieber, D., Singh, R., Shojaei, H., Ho, R, Sutton, C, : Learning Semantic Representations to Verify Hardware Designs. *Proceedings of the 35th Annual Conference on Neural Information Processing Systems (NIPS'21),* 23491–23504

235. Vaswani, A., Shazeer, N., Parmar, N., Uszkoreit, J., Jones, L., Gomez, A.N., Kaiser, L., Polosukhin, I.: Attention is all you need. *Proceedings of the 30th Annual Conference on Neural Information Processing Systems (NIPS'17),* 5998–6008

236. Verma, S., Dickerson, J., Hines, K.: Counterfactual explanations for machine learning: a review. https://arxiv.org/pdf/2010.10596.pdf

237. Wang, D., Zhang, Z., Zhang, H., Qian, Z., Krishnamurthy, S. V., Abu-Ghazaleh, N.: SyzVegas: Beating kernel fuzzing odds with reinforcement learning. *Proceedings of the 30th USENIX Security Symposium (USENIX Security'21),* 609–626

238. Wang, J., Song, C., Yin, H.: Reinforcement learning-based hierarchical seed scheduling for greybox fuzzing. *Proceedings of the 28th Network and Distributed System Security Symposium (NDSS'21)*

239. Wang, M., Deng, W.: Deep visual domain adaptation: A survey. *Neurocomputing,* 312:135–153 (2018)

240. Wang, S., Wang, P., Wu, D.: Semantics-aware machine learning for function recognition in binary code. *Proceedings of the 2017 IEEE International Conference on Software Maintenance and Evolution (ICSME'17),* 388–398

241. Wang, S., Wang, P., Wu, D.: Reassembleable disassembling. *Proceedings of the 25nd USENIX Security Symposium (USENIX'15)*

242. Wang, J., Shi, D., and Li, Y., Chen, J., Ding, H., and Duan, X.: Distributed framework for detecting PMU data manipulation attacks with deep autoencoders. *IEEE Transactions on Smart Grid,* 10(4):4401–4410 (2018)

243. Wang, X., Qin, Y., Wang, Y., Xiang, S., Chen, H.. ReLTanh: An activation function with vanishing gradient resistance for SAE-based DNNs and its application to rotating machinery fault diagnosis. *Neurocomputing* 363: 88–98 (2019)

244. Wei, S., Heng, Y., Chang, L., Dawn, S.: DeepMem: learning graph neural network models for fast and robust memory forensic analysis. *Proceedings of the 25th ACM SIGSAC Conference on Computer and Communications Security (CCS'18),* 606–618

245. Wiering, M., van Otterlo, M. (eds): Reinforcement learning state-of-the-art. Springer, 2012

246. Woods, J.: Machine Learning Robustness: New Challenges and Approaches. Vector Institute, 2022, https://vectorinstitute.ai/2022/03/29/machine-learning-robustness-new-challenges-and-approaches/

247. Xie, N., Zeng, F., Qin, X., Zhang, Y., Zhou, M., Lv, C.: Repassdroid: Automatic detection of android malware based on essential permissions and semantic features of sensitive apis. *Proceedings of the 16th IEEE International Symposium on Theoretical Aspects of Software Engineering (TASE'18),* 52–59

248. Xu, Z., Stoller, S. D.: Mining attribute-based access control policies. *IEEE Transactions on Dependable and Secure Computing* 12(5):533–45 (2014)

249. Yan, J., Yan, G., Jin, D.: Classifying malware represented as control flow graphs using deep graph convolutional neural network. *Proceedings of the 49th IEEE/IFIP International Conference on Dependable Systems and Networks (DSN'19),* 52–63

250. Yang, Q., Liu, Y., Cheng, Y., Kang, Y., Chen, T., Yu, H.: Federated learning. *Synthesis Lectures on Artificial Intelligence and Machine Learning*, 13(3):1–207 (2019)

251. Yang, S., Chakraborty, P., and Bhunia, S: Side-channel analysis for hardware Trojan detection using machine learning. *Proceedings of the 5th IEEE International Test Conference India (ITC India'21)*, 1–6

252. Zalewski, M.: American Fuzzy Lop. https://lcamtuf.coredump.cx/afl/

253. Zhang, J., Gao, C., Gong, L., Gu, Z., Man, D., Yang, W., Li, W.: Malware detection based on multi-level and dynamic multi-feature using ensemble learning at hypervisor. *Mobile Networks and Applications*, 26(4), 1668–1685, 2021.

254. Zhang, Z., Mudgerikar, A., Singla, A., Leung, K., Bertino, E., Chan, K., Melrose, J., Tucker, J.: Reinforcement and transfer learning for distributed analytics in fragmented software defined coalitions. *Proceedings of SPIE 11746, Artificial Intelligence and Machine Learning for Multi-Domain Operations Applications III, 117461W (April 2021)*

255. Zhao, Hang and Wang, Yujing and Duan, Juanyong and Huang, Congrui and Cao, Defu Tong, Yunhai and Xu, Bixiong and Bai, Jing and Tong, Jie Zhang, Qi: Multivariate time-series anomaly detection via graph attention network. *Proceedings of the 20th IEEE International Conference on Data Mining (ICDM'20)*, 841–850

256. Zhu, K., Chen, Z., Peng, Y., Zhang, L.: Mobile edge assisted literal multi-dimensional anomaly detection of in-vehicle network using LSTM, *IEEE Transactions on Vehicular Technology*, 68(5):4275–4284 (2019)

257. Zong, B., Song, Q., Min, M. R., Cheng, W., Lumezanu, C., Cho, D., Chen, H.: Deep autoencoding Gaussian mixture model for unsupervised anomaly detection. *Proceedings of 6th the International conference on learning representations (ICLR'18)*, poster paper

258. Zytek, A., Arnaldo, I., Liu, D., Berti-Équille, L., Veeramachaneni, K.: The need for interpretable features: motivation and taxonomy. *SIGKDD Explorations*, 24(1): 1–13 (2022)